Scorpions

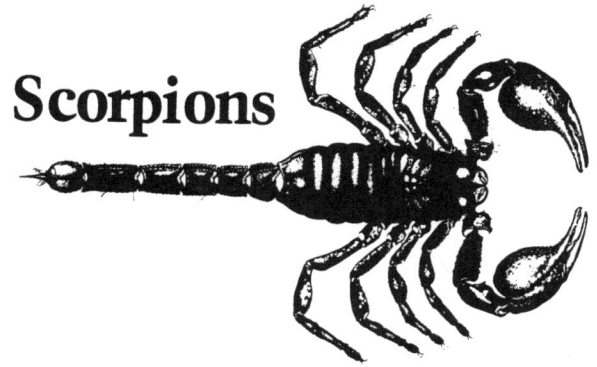

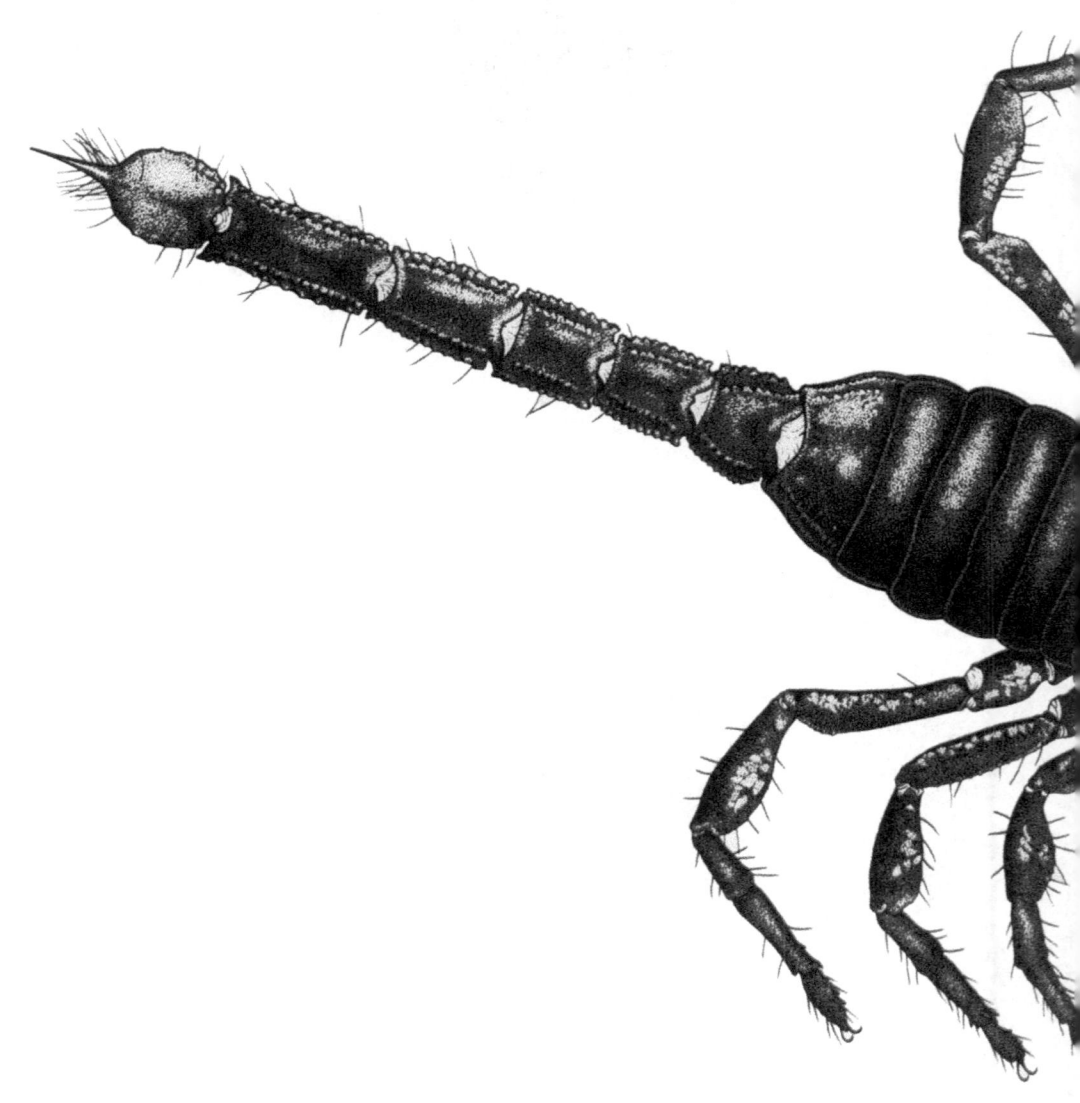

Medical Importance

by Hugh L. Keegan

UNIVERSITY PRESS
OF MISSISSIPPI • JACKSON

Library of Congress Cataloging in Publication Data

Keegan, Hugh L
    Scorpions of medical importance.
    Bibliography: p.
    Includes indexes.
    1. Scorpions. 2. Scorpions—Venom—Toxicology.
3. Antivenins. I. Title. [DNLM: 1. Scorpions.
QX469 K26s]
QL458.7.K44   615.9'42   80-16419

ISBN: 978-1-60473-378-5

Contents

Acknowledgments

IN ADDITION to those individuals whose assistance is specifically acknowledged in the various chapters of the text, a large number of persons in different walks of life helped in many ways to bring this project to a conclusion. I would first like to express my appreciation to the group of talented and dedicated Japanese artists formerly employed at the 406th Medical General Laboratory, U.S. Army Medical Command, Japan, who prepared the outstanding drawings used in the illustrations of species. These were: Mr. K. Daishoji, Mr. S. Ohtawa, Mr. S. Fujisawa, Mr. T. Ando, and Mrs. K. Miyasaka. Miss Yoshiko Yoshida, preparator for the department, was also of great assistance. Although the illustrations were completed during assignment of the writer to the 406th Medical General Laboratory during the period 1962–1966, the artists worked under the capable guidance of Major Gordon Field, chief of the Department of Entomology of the laboratory from 1960 until September 1962.

Specimens for study and illustration were obtained through the kindness of scientists at various institutions in the United States and overseas. Examples of several species were loaned from the collections of the United States National Museum by Dr. Ralph E. Crabill of the Division of Insects. Material from Egypt was obtained through the courtesy of Dr. Harry Hoogstraal of the U. S. Naval Medical Research Unit No. 3 in Cairo. Additional African scorpions were loaned by Dr. R. F. Lawrence of the Natal Museum, Pietermaritzburg, Natal, South Africa, and Dr. A. J. Hesse of the South African Museum, Capetown. South American specimens were supplied through the kindness of the late Dr. Flavio Fonseca, director of the Instituto Butantan, São Paulo, and his colleague Dr. Wolfgang Bücherl.

Arrangements for collection of specimens in Mexico were made with the help of the late Dr. Louis Mazzotti, Institute of Health and Tropical Diseases, Mexico, D. F.; Dr. Angel de la Garza Brito, director of the National Institute of Hygiene, Mexico, D. F., and Dr. Enrique Antonio Voges Herrera of the Secretariat of Health and Welfare, Colima, Colima. Public Health officials in Durango, Tepic, Manzanillo, and Mazatlan were uniformly most courteous and highly cooperative during our field work in their areas.

The U.S. Army Medical Research and Development Command, Office of the Surgeon General, was the primary source of financial support for this project. However, the manuscript could not have been brought to completion without the assistance, financial and otherwise, of the University of Mississippi Medical Center and particularly members of the Department of Preventive Medicine. Dr. Thomas J. Brooks, Jr., chairman of the department, was particularly helpful in his critical review of the manuscript. During the past several years, a series of departmental secretaries struggled with unfamiliar technical terms as the manuscript made its way through several drafts. The latest of these, Mrs. John Kinser, is typical of the group in her patience and cheerfulness in what must have been an arduous task.

Finally, I wish to express my appreciation to four friends who served with me in the U.S. Army at Ft. Sam Houston, Texas and who were invaluable not only in field collecting in Texas, Arizona, and Mexico, but also in maintenance of scorpion colonies in the laboratory. These are: Dr. (then Colonel) F. W. Whittemore, and three non-commissioned officers— Clevis M. Fitzgerald, Herman A. Bryant, and James F. Flanigan. I will never forget their skill, dedication and optimism, often under rather difficult circumstances.

Preface

THIS PUBLICATION is an account of the distribution, morphology, biology and classification of the scorpions considered to be of public health importance. Information is also given on clinical aspects of envenomation by scorpions and on methods for scorpion control. Judgment of the medical importance of a scorpion species has been based primarily on papers in professional journals and monographs, though correspondence with public health officials throughout the world and the personal field experience of the author in areas where scorpion sting is a matter of concern have also been used. This literature search was continued through the spring of 1978.

The discussion on general aspects of scorpion morphology and the descriptions of individual species were not written with the professional entomologist or arachnologist in mind. In several of the genera discussed here, only a few species are of medical interest. In such cases, no attempt has been made to describe or even mention those species lacking in medical importance. However, pertinent references on these topics, which might be of interest to the specialist, are given at the end of each chapter.

Finally, while there is a growing interest in the biochemistry and pharmacology of scorpion venoms these subjects are, by intent, not covered here. The sheer volume of published reports in these fields is such that a review of results obtained would increase the length of this publication to an impractical extent.

For those interested, the most recent and authoritative reviews of research on the nature of scorpion venoms are given in Chapters 12-15 (pp. 277–418) in Arthropod Venoms, S. Bettini (Ed.), 1978. Vol 48 in the New Series of the Handbook of Experimental Pharmacology. Springer-Verlag. Berlin, Heidelberg, New York.

Hugh L. Keegan

Scorpions

2

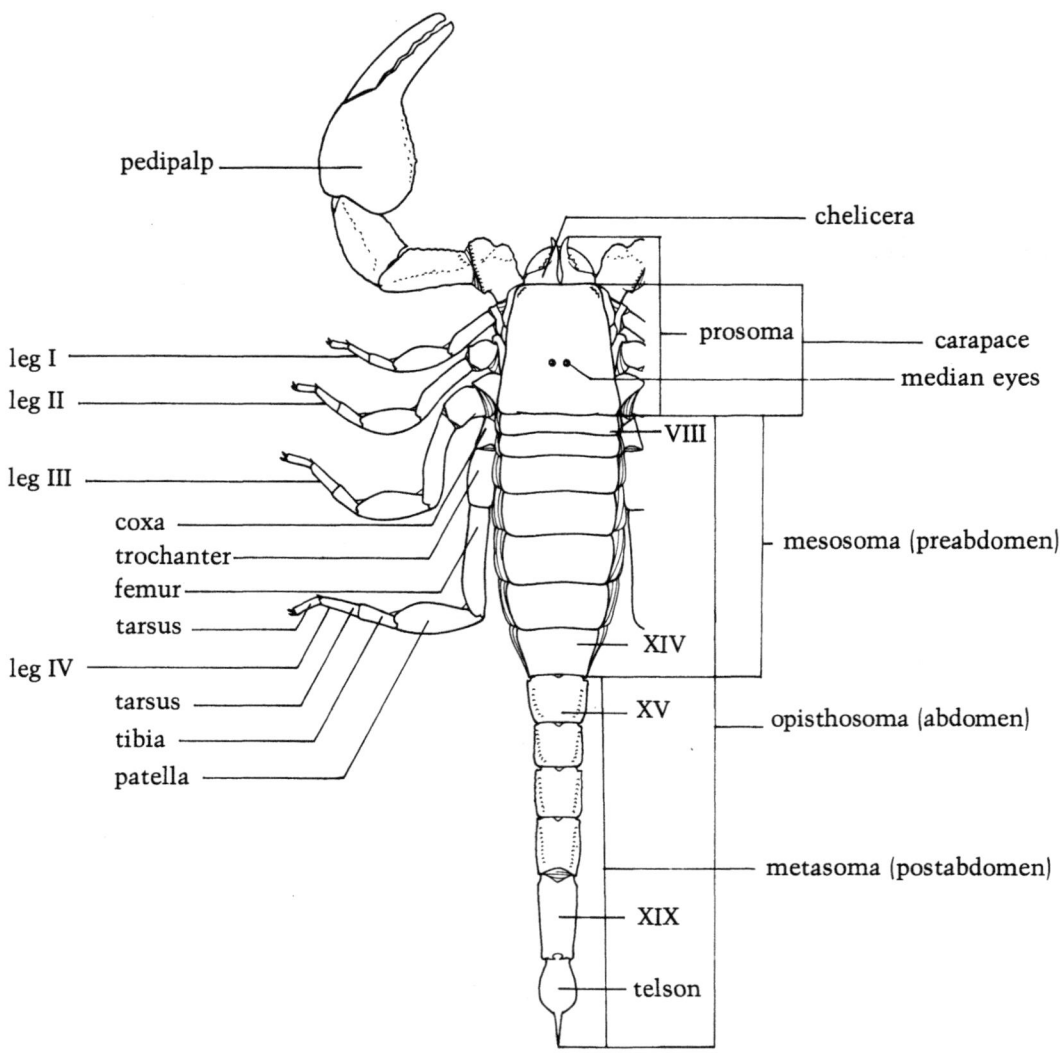

pedipalp

chelicera

prosoma — carapace
median eyes

leg I
leg II

VIII

leg III

coxa
trochanter
femur
tarsus

mesosoma (preabdomen)

leg IV

tarsus
tibia
patella

XIV

XV

opisthosoma (abdomen)

metasoma (postabdomen)

XIX

telson

PLATE I   Scorpion Morphology:
Dorsal View of Adult

1

Scorpion Morphology and Biology

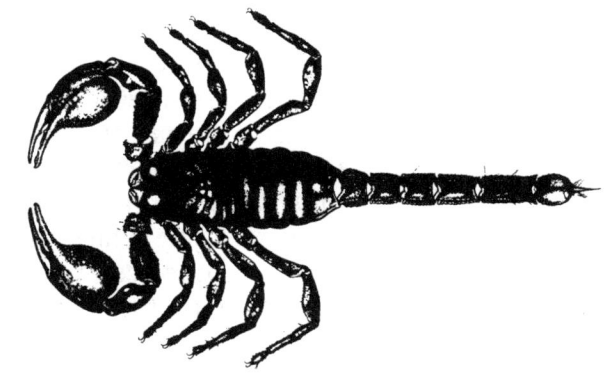

ALTHOUGH A DETAILED ACCOUNT of the internal and external morphology of scorpions is not given in this publication, those characteristics most commonly used in keys for identification of scorpions are defined in the following discussion and illustrated in plates I–V and in the drawings of individual species.

1. *General:* A scorpion is an arachnid with two main body divisions: the *prosoma* or *cephalothorax* and the *opisthosoma.* The latter is sometimes referred to as the *abdomen* and is further subdivided into an anterior *mesosoma* (*preabdomen*) and a posterior *metasoma* (*postabdomen*). Both of these regions are segmented. The narrow, tail-like metasoma, which in some keys is referred to as the *cauda,* terminates in the *telson.* This structure, which is vesicular in shape, possesses a sharp, curved spine or *aculeus.* On either side of the aculeus, near its tip, is a tiny pore which serves as an outlet for venom secreted by each of the two venom glands contained in the telson. Appendages of the scorpion are: the conspicuous, pincer-like *pedipalps;* the small, inconspicuous *chelicerae* between the bases of the pedipalps; four pairs of legs, all on the prosoma; and a pair of ventrally located, comb-like structures called *pectines* which have a sensory function. There are four pairs of *spiracles* on the venter of the mesosoma. These apertures of the respiratory organs, the book lungs, are either slits or oval in shape.

2. *Prosoma:* The prosoma is covered dorsally by a shield called the *carapace.* The surface of the carapace is marked by *transverse grooves* or *furrows, lateral depressions,* and a *median groove* (or furrow) which extends from the eyes to the

3

posterior margin of the carapace where it widens to form a *triangular depression. Keels* are usually evident on the carapace. These are called the *fore, middle, hind* and *lateral keels* and are often useful in classifying specimens. That portion of the *fore median keel* which is above a median eye is called a *superciliary ridge.* There is a pair of *median eyes* on an *ocular tubercle* and two groups of more anteriorly situated *lateral eyes.* The total number of eye pairs may vary from none to six, as some scorpions are eyeless. The taxonomically important structures on the ventral surface of the prosoma are the *coxae* of the legs and a small median plate, the *sternum.* The sternum may be triangular or pentagonal in outline, or may be divided transversely in some species. It is important in separating the various families of the scorpion.

3. *Mesosoma:* The dorsum of the mesosoma is distinctly segmented. Each of the seven segments is covered by a *tergal plate* (or shield). The ventral surface of the mesosoma is divided into four *sternites,* each with a pair of slit shaped *spiracles.* The genital aperture, the *gonotreme,* is between the coxae of leg pair IV. It is covered either by a pair of small plates or by a single plate, the *operculum,* and is bordered posteriorly by a transverse *postgenital fold.*

4. *Metasoma:* The metasoma or "tail" consists of six segments (somites) plus the telson. The first segment is conical with a tergite and a sternite which are trapezoidal in shape. This first segment is followed by five ringlike segments. The dorsal, lateral, and ventral surfaces of each segment may bear *keels* which in some cases may be armed with small spines and/or granules. The two pairs of keels nearest the dorsal medline are called the *dorsal median keels.* Those in a corresponding position on the ventral surface are the *ventral median keels. Dorsolateral* and *ventrolateral* keels may also be present. Although keels are not found in all species of scorpions, these structures are important aids in identification. A short *accessory* spine may be present ventral to the main spine of the telson. The presence or absence of this structure and its configuration are important in scorpion classification.

Two pyriform venom glands are contained in the telson. Each gland discharges its contents through a duct leading to a pore on the curved sting. Venom expulsion is accomplished through contraction of thick muscle layers that, investing each gland dorsally and mesally, compress the gland against the cuticle of the telson. Venom glands of scorpions show differences in internal structure that are of generic and, in

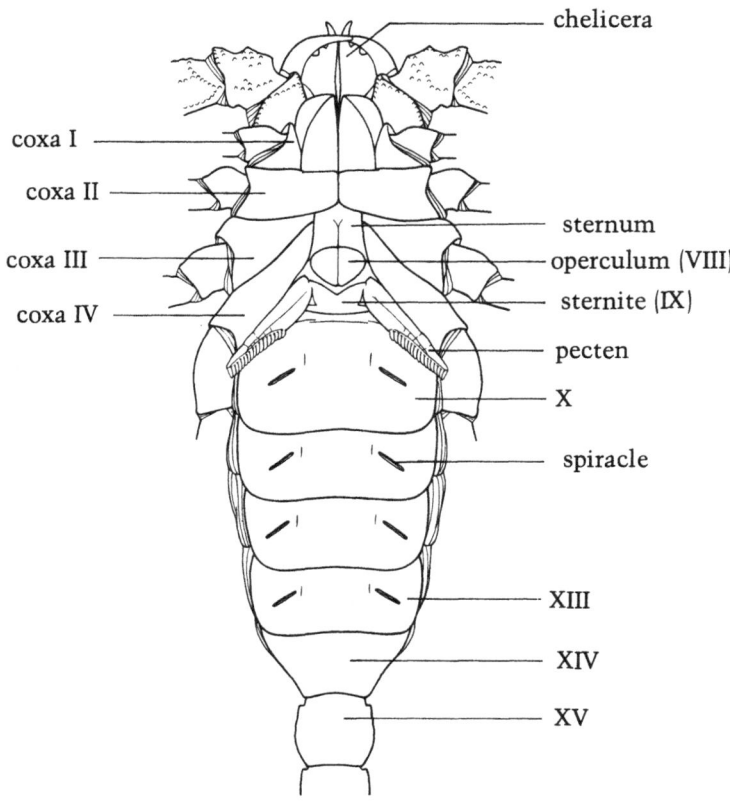

PLATE II   Scorpion Morphology:
            Ventral View of Adult

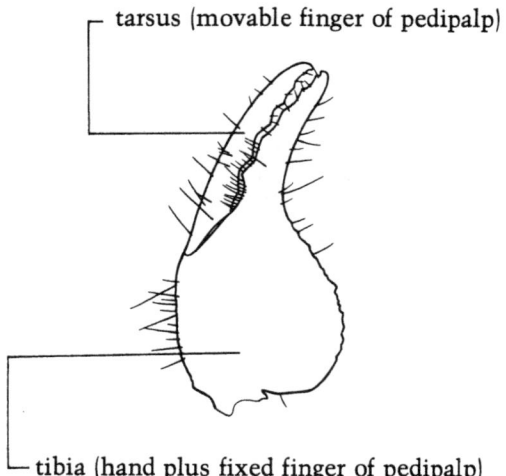

a. Tibia and tarsus of pedipalp

c. Genital region and pectines

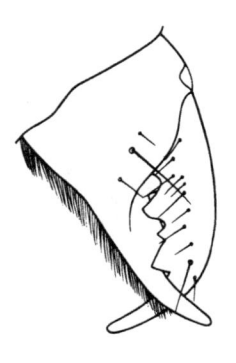

b. Chelicera

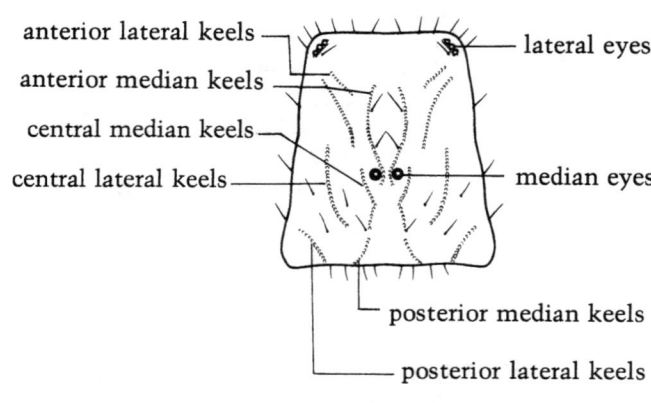

d. Carapace

PLATE III   Scorpion Morphology:
Additional Structures of Taxonomic Importance

some instances, of family status. These variations are due to the nature of the secretory epithelium. In some species the gland is a simple, sack-like structure, in others the epithelium may be thrown into numerous ridges, folds and irregular finger-like processes which greatly increase the secretory surface. The matter has been summarized by Junqua and Vachon (1968). Both light and electron microscopy were utilized by Keegan and Lockwood (1971) in examination of the secretory epithelium in venom glands of two species of scorpion, *Centruroides vittatus* (Say) from Texas and *Centruroides limpidus tecomanus* (Hoffman) from Mexico. Although the species differ greatly in their toxicity for man, no differences in the morphology of their venom glands could be detected.

5. *Appendages:* Scorpions possess seven pairs of appendages. These are: the *chelicerae* (one pair), the *pedipalps* (one pair), the legs (four pairs) and the *pectines* (one pair). All of these except the pectines are borne on the prosoma. The latter are located on the ventral surface of the mesasoma. The chelicerae consist of three segments. These are: the *basal portion*, a *movable finger* and a *fixed finger*. The number, relative size, and arrangement of the teeth on these fingers are useful in classification. The *pedipalps* are the large, conspicuous "pincers" of the scorpion. Each pedipalp consists of six segments: *coxa, trochanter, femur, patella, tibia,* and *tarsus*. The basal (proximal) portion of the tibia is called the "hand." The size, location and number of teeth, as well as various keels, granules, tubercles and lobes present on the "hand" are of taxonomic importance. The number and arrangement of sensory bristles called *trichobothria* which arise from cup-like areolae on the femur, patella, and tibia of the pedipalps are also of considerable importance in scorpion taxonomy. The *legs* consist of seven segments. These are: coxa, trochanter, femur, patella, tibia, tarsus and *pretarsus*. The pretarsus bears two *lateral claws* and a *median claw*. The tarsus itself is divided into *tarsomeres* I and II. Comparative measurements of the various leg segments are useful in classification. Also, variations in size and shape of *tibial spurs, pedal spurs* (found in all scorpions between tarsomeres I and II) the *median* and *lateral lobes* of tarsomere II, and the length of the median claw of the pretarsus are used in keys for identification of scorpions.

Determination of the sex of a scorpion may be difficult. While in some taxa segments of the cauda of the female are relatively shorter than in the male, there is no single external

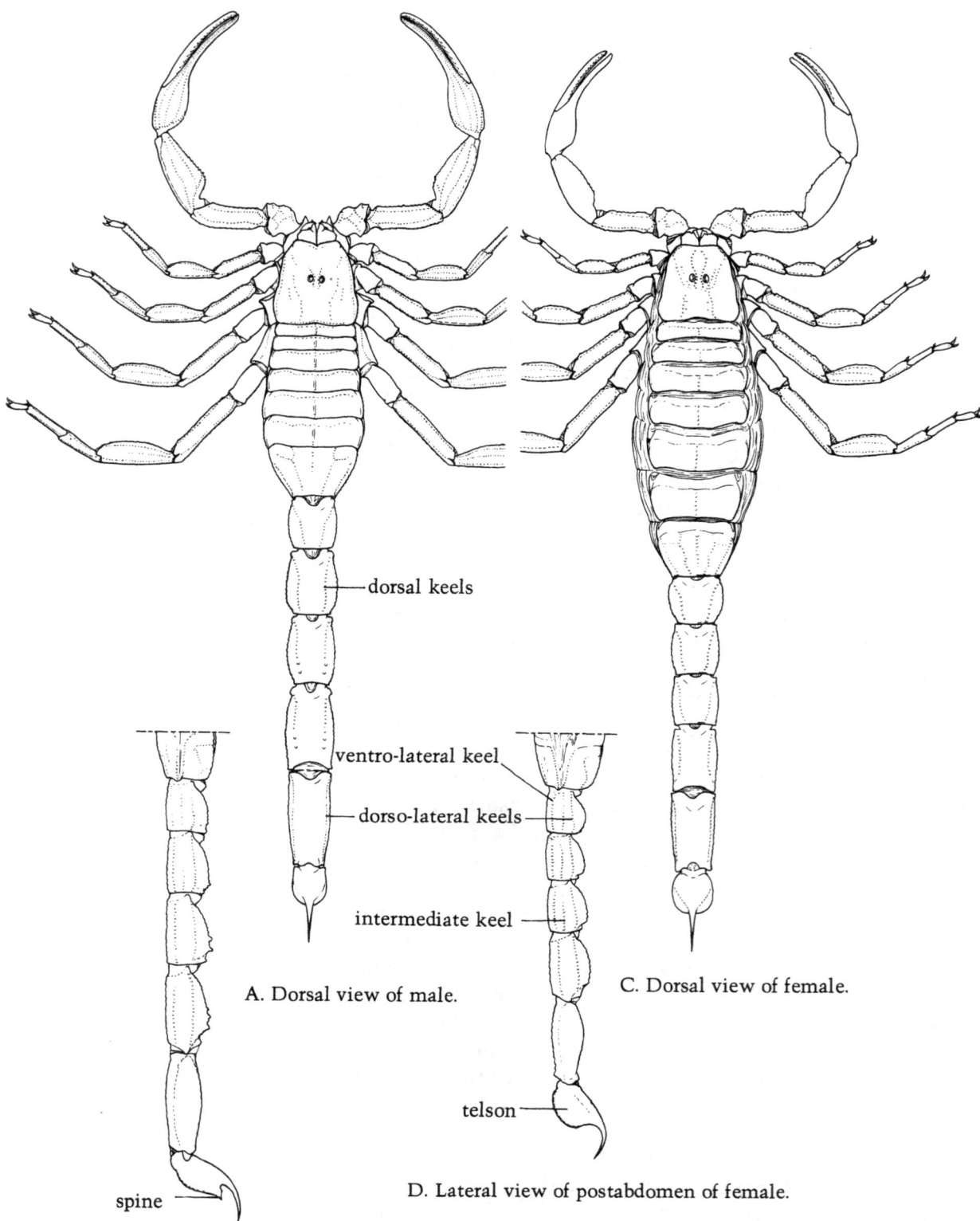

dorsal keels

ventro-lateral keel

dorso-lateral keels

intermediate keel

A. Dorsal view of male.

C. Dorsal view of female.

telson

spine

D. Lateral view of postabdomen of female.

B. Lateral view of postabdomen of male.

PLATE IV   Scorpion Morphology:
Sexual Dimorphism in the Indian Scorpion *Buthotus tamulus*

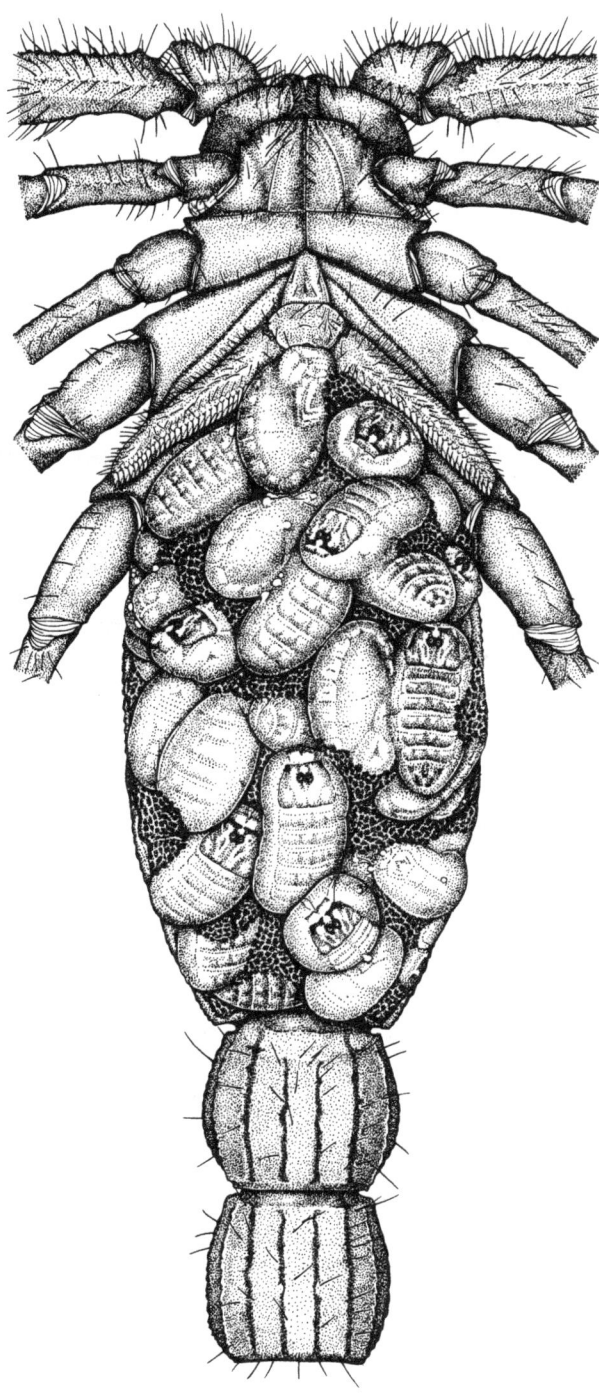

PLATE V Scorpion Morphology:
A Gravid Female of *Buthotus tamulus*

characteristic common to all scorpions to allow easy differentiation of the sexes.

While some scorpions such as species of the African genus *Pandinus* may attain a length of 200 mm or more, most are much smaller. However, size alone is not a valid criterion for judgment of the hazard posed by a scorpion. For example, the dangerously venomous species of genus *Centruroides* of the New World seldom reach a length greater than 100 mm.

More detailed information on scorpion morphology is available in publications by Kraepelin (1899), Werner (1935), Snodgrass (1952), Comstock (1940), Vachon (1952 and 1973) and most recently by Stahnke (1970).

Although the general aspects of scorpion biology are well known, little or no information is available concerning many species of medical importance. Scorpions are nocturnal animals. During the daytime they remain hidden and emerge at night to feed. Some scorpions live in burrows, others hide under rocks, logs, or in crevices. The last mentioned are probably responsible for most cases of scorpion sting in man, for as they seek refuge when daylight comes, they hide in shoes, blankets or in clothing which has been left upon the floor. This is a problem of particular importance in areas where housing construction is poor, screens are lacking, and cracks in the walls or floor offer many hiding places for scorpions.

Food consists of insects and other arthropods of suitable size although cannibalism may also occur, and examples of some of the larger species have been seen feeding on lizards, snakes and other small vertebrates. Whittemore *et al* (1963) and Deoras (1961) found that laboratory colonies could be easily maintained on a diet of crickets and roaches. Bücherl (1971) found that specimens of the dangerously venomous *Tityus serrulatus* would feed readily on fragments of freshly killed tarantulas. The prey is captured by the pedipalps and if it is large or active it is stung repeatedly as the postabdomen or "tail" is arched over the back and the spine at the end of the telson is brought into action. After the prey has been captured by the pedipalps, it is moved to the chelicerae where a process of maceration begins. As the chelicerae tear the food into tiny particles, these are packed between the coxae of the pedipalps. At the same time, a secretion which liquifies tissues of the prey flows from the buccal area of the scorpion. As this happens, the coxae compress the food particles, and the "juice" thus produced is ingested by the scorpion. When all fluid has been expressed from food particles, the remaining dry pellets are discarded. The process has been described in detail by Stahnke (1966). Although scorpions will feed often if food is available, they, like the spiders, can survive for long periods of time without food if water is available.

It is apparent that the eyes of the scorpion play only a minor role in location of prey. This is accomplished primarily by the pectines, the trichobothria and other hairs on the appendages and body. While a variety of functions has been credited to the pectines, not all of these have been substantiated. Stahnke (1966), who reviewed literature on the subject, found that in his studies of *Centruroides sculpturatus* the pectines definitely appeared to function as chemoreceptors. He found that a hungry scorpion would crawl over a dead insect until its pectines made contact, it would then back up and pick up the prey. The trichobothria also are extremely sensitive structures. The slightest contact with a moving object will trigger an instant response by a hungry scorpion. Schultze (1927), in his excellent study on the biology of *Heterometrus longimanus*, noted that hairs on many parts of the body seemed to have a sensory function, and that a female specimen observed seemed able, by means of these, to distinguish between her own young and small insects which had been put into the cage as food.

Scorpions produce living young. Plate 5 shows the appearance of well developed larvae within the female. In some species, the larvae are born in a transparent birth membrane from which they quickly emerge. In others, the young have already emerged from the membrane at the time of birth. During birth of the young, many female scorpions fold one or more of the anterior pairs of legs under the genital operculum forming what has been called a "birth basket" which "catches" the emerging young so that they do not touch the ground. Shortly after birth and emergence from the birth membrane, the young scorpions climb up one of the legs onto the back of the mother. Williams (1969) found that in species of family Vejovidae studied by him, the young scorpions soon assumed a definite nonrandom orientation on the back of the mother. In contrast, larvae of species of family Buthidae exhibited a haphazard "pile-up" and were found randomly in layers three to four individuals deep.

Newly born scorpions are whitish in color and, in most instances, do not achieve adult coloration until at least after two or three moults. The young remain on the back of the mother until after the first moult which takes place from about a week to two weeks or more following birth. During this period they do not feed. Various writers (Vachon, 1970; Schultze, 1927; and Williams, 1969) have noted that all members of a litter experience the first moult almost simultaneously or at least within a period of about 24 hours. Following descent from the mother, the young scorpions scatter and commence to feed. Growth and the number and frequency of subsequent moults depend upon food supply, temperature, and other environmental factors. Because of this there is considerable variation in the time required to reach maturity. In their study of the large African scorpion, *Pandinus gambiensis,* Vachon *et al* (1970) found that these scorpions attained maturity over a period ranging from three years, seven months to seven years. Other writers, including Schultze (1927) and Stahnke (1966) have reported similar variation both in number of moults and in time required to reach maturity.

Note: Some writers refer to newly born scorpions as larvae, others call them first instar nymphs, and still others avoid technicalities by using the terms "juveniles" or "young." In the strict sense probably the term "nymph" is correct, for unlike the larval ticks and mites, these have the complete complement of four pairs of legs, as in the spiders.

Before insemination, scorpions conduct an elaborate courtship or "promenade a deux" during which the male grasps the pedipalps of the female with his own and "waltzes" back and forth, sometimes for hours. Alexander (1957 and 1959) and Rosin and Shulov (1963) considered that this performance is actually a search for a suitable substrate for deposition of a sperm container, the spermatophore, and that when this has been found the promenade ends. Upon reaching its objective, the male extrudes a spermatophore which is fastened to the substrate by an adhesive fluid. When this has been accomplished, the male maneuvers the female over the spermatophore so that a centrally located capsule containing spermatic fluid comes into contact with her genital aperture. Subsequently, the pressure caused by the bending of the spermatophore opens or ruptures the capsule, releasing the sperm mass into her genital tract.

The structure of the spermatophore, which is quite complex, varies among families of scorpion. Alexander (1957 and 1959) could well be described as the pioneer in the study of spermatophore structure. Rosin and Shulov (1963) presented a careful analysis of the spermatophore of the diplocentrid scorpion *Nebo hierochonticus*.

Following mating, at least in some species, the male may immediately attack the female, forcing her to retreat. This courtship and mating procedure have been described in precise detail by Alexander (1957 and 1959), Rosin and Shulov (1963), and in a charming literary style by that meticulous observer Fabre (1911). Williams (1969) gave an excellent account of the birth activities of some species of the four families of scorpion in the United States.

Parthenogenesis also occurs among scorpions. Bücherl (1971) wrote that he had never found a male of *Tityus serrulatus* among the approximately 60 thousand specimens kept in his laboratory over the years. This would appear to corroborate the earlier reports of parthenogensis in this species by Matthiesen (1962) and San Martin *et al* (1966). The extent to which this occurs among other species is not known.

There have been several papers on gametogenesis in scorpions in recent years. Most of these have dealt with species of the Indian genus *Heterometrus (Palamnaeus)*. Oustanding among these have been papers by Bedi (1962 and 1963), Lal Sareen (1962), Sharma *et al* (1962), Srivastava and Uma (1961) and Venkatanarasimhaiah and Rajasekarasetty (1964). Studies of chromosome numbers in male germ cells of scorpions have shown surprising variation, even among species of the same genus. Yoshida and Toshioka (1964) found that the haploid number of chromosomes in primary and secondary spermatocytes of the Indian scorpion, *Heterometrus gravimanus*, was 27. In *Buthotus tamulus*, another Indian species which they studied, the haploid number was 12. The diploid number of 24 was seen in spermatogonial cells near the testis wall. In the Texas striped scorpion, *Centruroides vittatus*, primary and secondary spermatocytes, without exception, showed 11 haploid chromosomes at metaphase. The diploid number was again observed in spermatogonial cells. As far as can be determined, information concerning chromosome numbers has not yet been utilized in attempts to determine relationships of taxa.

Regeneration of appendages, also found in other arthropod taxa, occurs in scorpions. Several writers, most recently Rosin and Shulov (1963), have reported this phenomenon. It was found that specimens of *Nebo hierochonticus*, a species widely distributed in the Middle East and North Africa could regenerate not only appendages, but also the aculeus or stinging spine at the end of the telson. In some instances, the regenerated part was incomplete or malformed.

Excellent general accounts which cover various aspects of scorpion biology have been written by Baerg (1961), Stahnke (1956 and 1966), Bücherl (1971), Ennik (1972), and Levi and Levi (1968).

References on Morphology and Biology

Alexander A. 1957. The courtship and mating of the scorpion *Opisthophthalmus latimanus. Proc. Zool. Soc. Lond., 128:* 529–545.

———. 1959. Courtship and mating in the buthid scorpions. *Proc. Zool. Soc. Lond., 133:* 145–169.

Baerg, W. J. 1961. Scorpions: biology and effect of their venom. *U. of Arkansas, Agr. Exp. Sta. Bull. No. 649,* 34 pp.

Bedi, U. 1962. Studies on the male germ cells of scorpions *Palamnaeus bengalensis* and *Palamnaeus fulvipes*, with particular reference to the morphology and cytochemistry of the cytoplasmic inclusions. *Res. Bull. Panjab Univ. Sci., 13:* 213–225.

———. 1963. "Chromotoid body" in the spermatogenesis of scorpions. *Experientia. 19:* 90–91.

Bücherl, W. 1971. Classification, Biology, and Venom Extraction of Scorpions. Ch. 55 *In: Venomous Animals and Their Venoms,* vol. III, *Venomous Invertebrates.* Bücherl, W. and Buckley, E. E., eds. Academic Press, New York, pp. 317–347.

Comstock, J. H. (Revised and edited by W. J. Gertsch, 1940). *The Spider Book.* Doubleday, Doran & Co., Inc., New York, 729 pp.

Deoras, P. J. 1961. A study of scorpions. Their distribution, incidence and control in Maharashtra. *Probe, 1:* 45–54.

Ennik, F. 1972. A short review of scorpion biology, management of stings, and control. *Calif. Vector Views, 19:* 69–79.

Fabre, J. H. 1911. Chapters XVII and XVIII on the Languedocian scorpion *In: The Life and Love of the Insect.* A. and E. Black, Ltd., London, pp. 223–260.

Junqua, C. and Vachon, M. 1968. *Les Arachnides Venimeus et Leurs Venins.* État actuel des recherches. Academie royale des Sciences d'Outre-Mer. Classe des Sciences naturelles et medicales, N. S. XVII-5. Brussells, 136 pp.

Keegan, H. L. and Lockwood, W. R. 1971. Secretory epethelium in venom glands of two species of scorpion of the genus *Centruroides* Marx. *Am. J. Trop. Med. and Hyg., 20:* 770–785.

Kraepelin, K. 1899. Scorpiones und Pedipalpi. Das Tierreich. Friedlander und Sohn, Berlin. 8 Lieferung, 265 pp.

Lal Sareen, M. 1962. Morphological and histological studies on the female germ cells of scorpions *Palamnaeus fulvipes* Koch and *Palamnaeus bengalensis* Koch. *Res. Bull. Panjab Univ. Sci.*, *13:* 71–83.

Levi, H. W. and Levi, L. R. 1968. *Spiders and their Kin.* Golden Press, New York, 160 pp.

Matthiesen, F. A. 1962. Parthenogenesis in scorpions. *Evolution, 16:* 255–256.

Rosin, R. and Shulov, A. 1963. Studies on the scorpion *Nebo hierochonticus. Proc. Zool. Soc. Lond., 140:* 547–575.

San Martin, P. R. and Gambardella, L. A. 1966. Nueva comprobacion de la partenogenesis en *Tityus serrulatus* Lutz y Mello-Campos 1922 (Scorpionida, Buthidae). *Rev. Soc. Ent. Arg., 28:* 79–84.

Schultze, W. 1927. Biology of the large Philippine forest scorpion. *Philip. J. Sci., 32:* 375–390.

Sharma, G. P., Ram. P., and Rajander, H. 1962. Meiosis in two species of *Palamnaeus* (Scorpiones-Scorpionidae). *Res. Bull. Panjab. Univ. Sci. 13:* 85–89.

Snodgrass, R. E. 1952. The Arachnida. Ch. V *In: A Textbook of Arthropod Anatomy.* Copyright 1952 by Cornell University. Reprinted by Arrangement. Hafner Publishing Co., New York, 1965, pp. 59–127.

Srivastava, M. D. L., and Uma, A. 1961. Absence of chiasmata and formation of a complex chromosomal body in the spermatogenesis of the scorpion *Palamnaeus longimanus. Caryologia, 14:* 63–77.

Stahnke, H. L. 1956. Scorpions. 2nd. Ed. Poisonous Animals Research Laboratory, Arizona State College, Tempe, Arizona, 36 pp.

———. 1966. Some aspects of scorpion behavior. *Bull. Southern Calif. Acad. Sci., 65:* 65–80.

———. 1970. Scorpion nomenclature and mensuration. *Ent News, 81:* 297–316.

Vachon, M. 1952. *Etudes sur Les Scorpions.* Institut Pasteur d'Algerie, Alger, 482 pp.

Vachon, M., Roy, R. and Condamin, M. 1970. Le developpement post-embryonnaire du scorpion *Pandinus gambiensis* Pocock. *Bull. IFAN, ser. A, 32:* 412–432.

Vachon, M. 1973. Étude des caracteres utilisés pour classer les familles et les genres de Scorpions (Arachnides). Bull. Mus. Hist. Nat., Paris 3e., No. *140* (Zoologie 104): 857–958.

Venkatanarasimhaiah, C. B. and Rajasekarasetty, M. R. 1964. Contribution to the cytology of Indian scorpions. Chromosomal behavior in the male meiosis of *Palamnaeus gravimanus. Caryologia. 17:* 195–201.

Werner, F. 1935. Scorpiones, Pedipalpi. *Bronn's Klassen und Ordnungen des Tierreichs,* 5 Band, IV abt., 8 Buch, 318 pp.

Whittemore, F. W., Keegan, H. L., Fitzgerald, C. M., Bryant, H. A. and Flanigan, J. F. 1963. Studies of scorpion antivenins. 2. Venom collection and scorpion colony maintenance. *Bull. Wld. Hlth. Org., 28:* 505–511.

Williams, S. C. 1969. Birth activities of some North American scorpions. *Proc. Calif. Acad. Sci., Ser. 4, 37:* 1–24.

Yoshida, Y. and Toshioka, S. 1964. Studies on spermatogenesis in scorpions. *Arach. Soc. of E. Asia, 19:* 1–4.

16

2

Geographic Distribution of Dangerously Venomous Scorpions

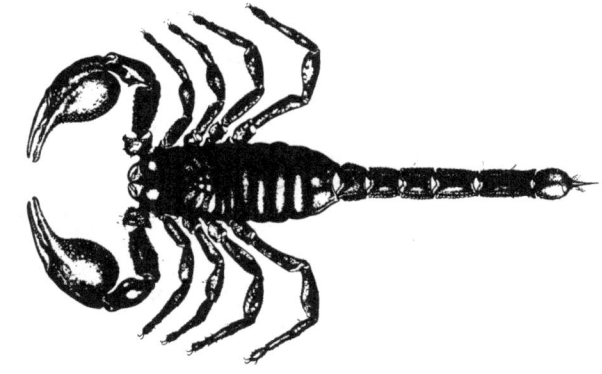

IN 1968, Junqua and Vachon compiled a list of 79 species of scorpion reputed to be of medical importance in that their stings often result in serious illness or death. However, more recent information indicates that some of the species listed may have been overrated and that others of true medical importance were not mentioned. Species in the following list are included because of published accounts, or as a result of personal correspondence with public health officials and other scientists. A major problem has been that data are lacking concerning the importance of many species in all countries where they may occur. While it might be assumed that a species of medical importance in one area would be of equal importance throughout its range, the area's life styles, economic development, housing conditions and effectiveness and availability of public health services may all affect the matter. For these reasons, listings given here for specific countries are based on availability of verifiable data. At the end of the listing for each major geographic area, information is given concerning those countries in which scorpion sting does not seem to be of medical importance.

Africa and the Middle East	
Species	*Sources of Information*
Algeria *Buthus occitanus*	Bouisset and Larrouy, 1962
Androctonus australis (L.) *hector*	Irunberry and Pilo-Moron, 1965
Egypt *Androctonus australis*	Vachon, 1966; Balozet, 1971
Androctonus amoreuxi	Vachon, 1966; Balozet, 1971
Leiurus quinquestriatus	Junqua and Vachon, 1968; Vachon, 1966; Balozet, 1971
Iraq *Hemiscorpion lepturus*	Pringle, 1960
Androctonus crassicauda	Pringle, 1965
Israel *Leiurus quinquestriatus*	Shulov, 1962; Gueron and Yaron, 1970
Androctonus crassicauda	Shulov, 1962; Gueron and Yaron, 1970
Androctonus bicolor	Shulov, 1962; Gueron and Yaron, 1970
Jordan *Buthus occitanus*	Wahbeh, 1965
Morocco *Androctonus mauritanicus*	Marcenac, 1926
Androctonus australis	Kupka, 1965
Androctonus amoreuxi	Levy, 1965
Buthus occitanus	Levy, 1965
South Africa (Union of) *Parabuthus triradulatus*	Mason, 1965
Parabuthus transvaalensis	Mason, 1965
Parabuthus villosus	Mason, 1965

NOTE: Although species of *Hadogenes, Uroplectes,* and *Opisthophthalmus* also are found in South Africa, an antivenin is prepared only against *Parabuthus* venom, as venoms of the others are not highly toxic.

Africa and the Middle East

Species	Sources of Information	
Buthus minax (L. Koch)	El-Asmar *et al*, 1974	**Sudan**

NOTE: While numbers of cases were not given, the authors indicated that the venom of this species, which is widely distributed in Sudan, was comparable in toxicity with that of *Leiurus quinquestriatus*. Hazard is due to the habit of scorpions of hiding under bark of trees used commonly as firewood. No data are available on symptoms in man or effectiveness of treatment.

Androctonus crassicauda	Tulga, 1964	**Turkey**
Leiurus quinquestriatus	Tulga, 1964	
Mesobuthus gibbosus	Tulga, 1964	

Countries in Africa and the Middle East in which Scorpion Sting Does Not Seem to be a Public Health Problem

Country	Sources of Information	
Zaire	Taufflieb, 1965	
Nigeria	Conrad, 1965	
Chad	Seymour, 1965	

NOTE: Dr. Seymour had heard of no deaths from scorpion sting in Chad but did mention that the sting of one small scorpion found there caused pain and could incapacitate an adult for several hours.

Tanzania	Pringle, 1965	
Kenya	Pringle, 1965	
Uganda	Pringle, 1965	
Buthotus tamulus	Deoras, 1961; Mundle, 1961; Santhanakrishnan & Balagopal Raju, 1974	**India**

Countries of Southeast Asia in which Scorpion Sting Does Not Seem to be a Public Health Problem

Country	Sources of Information
Thailand	Wongsarojama, 1965
Indonesia	Kopstein, 1927

NOTE: Although numerous species of scorpion genus *Heterometrus* occur in S. E. Asia, little is known concerning the venom of most of these. The few which have been studied are not dangerously venomous.

Mexico

Species	Sources of Information
Centruroides noxius	Hoffman, 1936; Diaz Najera, 1966; Whittemore and Keegan, 1963
Centruroides suffussus suffussus	Hoffman, 1936; Diaz Najera, 1966; Whittemore and Keegan,1963
Centruroides infamatus infamatus	Hoffman, 1936; Diaz Najera, 1966
Centruroides elegans	Hoffman, 1936; Diaz Najera, 1966
Centruroides limpidus limpidus	Hoffman, 1936; Diaz Najera, 1966
Centruroides limpidus tecomanus	Hoffman, 1936; Diaz Najera, 1966; Whittemore and Keegan, 1963

West Indies

	Species	Sources of Information
Trinidad	Tityus trinitatis	Waterman, 1938a, 1938b, 1950a, 1950b; Poon-King, 1963; Aitken, 1965

Central America

There are no data to indicate that scorpion sting is a problem of public health importance in Central America. Thaeler (1965) and Canton (1965) both indicated that scorpion sting was definitely not of significant medical importance in Nicaragua.

South America

Species	Sources of Information	
Tityus bahiensis	Abalos, 1963	**Argentina**
Tityus serrulatus	Bücherl, 1971	**Brazil**
Tityus bahiensis	Bücherl, 1971	
Tityus cambridgei	Floch, Barrat and Abonnenc, 1950	**Guyana**
Centruroides gracilis	Berti, 1965	**Venezuela**
Tityus trinitatis	Bücherl, 1971	

Countries in South America in which Scorpion Sting Does Not Seem to be a Public Health Problem

Country	Sources of Information
Colombia	Marinkelle and Stahnke, 1965; Botero, 1965

NOTE: While the sting of *Centruroides margaritatus*, a common species in Colombia, causes pain and discomfort, it is not regarded as dangerously venomous.

Guyana	Giglioli, 1965

United States of America

Species	Sources of Information
Centruroides sculpturatus	Stahnke, 1956; Stahnke, 1971

References
on Geographic
Distribution

Abalos, J. W. 1963. Scorpions of Argentina. *In: Venomous and Poisonous Animals and Noxious Plants of the Pacific Region.* Keegan, H. L. and Macfarland, W. V., eds. Pergamon Press, New York, pp. 111–117.

Balozet, L. 1971. Scorpionism in the Old World. *In: Venomous Animals and Their Venoms,* vol. III, *Venomous Invertebrates.* Bücherl. W. and Buckley, E. E., eds. Academic Press, New York, London, pp. 349–371.

Bouisset, L. and Larrouy, G. 1962. Envenimation par *Scorpio maurus* et *Buthus occitanus* dans le département de Tlecen. *Bull. Soc. Path. Exot., 55:* 139–146.

Bücherl, W. 1971. Classification, Biology, and Venom Extraction of Scorpions, Ch. 55 *In: Venomous Animals and Their Venoms,* vol. III, *Venomous Invertebrates.* Bücherl, W. and Buckley, E. E., eds. Academic Press, New York, London, pp. 317–347.

Deoras, P. J. 1961. A study of scorpions. Their distribution, incidence, and control in Maharashtra. *Probe, 1:* 45–54.

Diaz Najera, A. 1966. Alacranes de la republica Mexicana. Clave para identificas especies de *Centrurus. Rev. Inv. Sal. Publ. Mexico, 26:* 109–123.

El-Asmar, M. F., Soliman, S. F., Ismail, M., and Osman, O. H. 1974. Glycemic effect of venom from the scorpion *Buthus minas* (L. Koch). *Toxicon, 12:* 249–251.

Floch, H., Barrat, R., and Abonnenc, E. 1950. L'envenimation par piqûre de Scorpions en Guyane francaise. *Inst. Past. et Territoire Inini. Publ., 219:* 1-4.

Gueron, M. and Yaron, R. 1970. Cardiovascular manifestations of severe scorpion sting. *Chest, 57:* 156–162.

Hoffman, C. A. 1936. La distribution geografica de los alacranes peligrosas en La Republica Mexicana. *Bol. Inst. Hig., 2:* 321–330.

Irunberry, J. and Pilo-Moron, E. 1965. Renforcement de l'action antigénique de venin de scorpion (*Androctonus australia*) pars divers adjuvants. *Annales de L'Institut Pasteur, 108:* 378–383.

Junqua, C. and Vachon, M. 1968. *Les Arachnides Venimeux et Leurs Venins.* Etat actuel des recherches. Académie royale des Sciences d'Outre-Mer. Classe des Sciences naturelles et médicales, N. S. XVII-5. Brussells, 136 pp.

Kopstein, F. 1927. The poison of the Javanese giant scorpion *Heterometrus cyaneus. Meded. Dienst. d. Volksgezondheid in Nederl. Indie., 3:* 1–10.(In Dutch)

Levy, M. 1965. Les envenimations par piqûres de scorpions à Marrakech. (Maroc). Thèse, Faculté de Médecine, Paris, 42 pp.

Marcenac, M. 1926. Arachnides, myriapodes et serpents de la region du Tadla (Maroc). *Bull. Soc. Path. Exot., 19:* 560–563.

Marinkelle, C. J. and Stahnke, H. L. 1965. Toxicological and clinical studies on *Centruroides margaritatus* (Gervais), a common scorpion in western Colombia. *J. Med. Ent., 2:* 197–199.

Mundle, P. M. 1961. Scorpion stings. *Brit. Med. J., 1:* 1042.

Poon-King, T. 1963. Myocarditis from scorpion stings. *Brit. Med. J., 1:* 374–377.

Pringle, G. 1960. Notes on the scorpions of Iraq. *Bull. Endem. Dis., 3:* 73–87.

Santhanakrishnan, B. R. and Balagopal Raju, V. 1974. Management of scorpion sting in children. *J. Trop. Med. Hyg., 77:* 133–135.

Shulov, A. 1962. On some Israeli scorpions. *Dapin Refuiim Folia Medica,* *21:* 31–34.

Stahnke, H. L. 1956. Scorpions. Poisonous Animals Research Laboratory. Arizona State College, Tempe, 36 pp.

———. 1971. Some observations of the genus *Centruroides* (Buthidae, Scorpionida). *Ent. News, 82:* 218–307.

Tulga, T. 1964. Scorpions found in Turkey and paraspecific action of an antivenin produced with the venom of the species: *Androctonus crassicauda. Turk. Hig. Tecr. Biyol. Derg., 24:* 153–155.

Vachon, M. 1966. Liste des scorpions connus en Égypte, Arabie, Israël, Liban, Syrie, Jordanie, Turquie, Irak, Iran. *Toxicon, 4:* 209–218.

Wahbeh, Y. 1965. Scorpion stings in children. *Jordan Med. J., 1:* 57–61.

Waterman, J. A. 1938a. Some notes on scorpion poisoning in Trinidad. *Trans. Roy. Soc. Trop. Med. and Hyg., 31:* 607–624.

———. 1938b. A few cases. *Caribbean Med. J., 1:* 119–120.

———. 1950a. Two cases of scorpion poisoning characterized by convulsions with electrocardiograms. *Caribbean Med. J., XII:* 127–129.

———. 1950b. Scorpions in the West Indies with special reference to *Tityus trinitatis.* Caribbean Med. J., XII: 167–177.

Whittemore, F. W. and Keegan, H. L. 1963. Medically important scorpions in the Pacific area. *In: Venomous and Poisonous Animals and Noxious Plants of the Pacific Region.* Keegan, H. L. and Macfarlane, W. V., eds. Pergamon Press, New York, pp. 107–110.

Personal Communications (All Received During 1965)

Aitken, T. H. G. University of the West Indies, Trinidad Regional Virus Laboratory.

Berti, A. L. Director de Malariologia y Saneamiento Ambiental, Ministerio de Sanidad y Assistencia Social, Caracas, Venezuela.

Botero, D. R. Head, Department of Microbiology and Parasitology, Faculty of Medicine, Medellin, Colombia.

Canton, J. A. Director General PUMAR. U.S. Aid Mission to Nicaragua, Managua, Nicaragua.

Conrad, J. L. 1700 Clifton Road, Atlanta, Georgia. (Harvard School of Public Health at the time he wrote.) *Re:* Nigeria.

Giglioli, G. Medical Advisor, British Guyana Sugar Producer's Association.

Kupka, K. WHO Medical Officer, Bureau de Representant au Maroc. B. P. 520 Rabat-Chellah, Morocco.

Mason, J. H. The South African Institute for Medical Research, P. O. Box 1038, Johannesburg, South Africa.

Pringle, G. Director, East African Institute of Malaria and Vector Borne Diseases, East African Common Services Organization, Amani, Tangi, Tanzania.

Seymour, D. W. Le Centre Médical, Baptist Mission, Koumra, République du Chad.

Taufflieb, R. Institut de Recherches, Scientifiques au Congo, Brazzaville, Congo.

Thaeler, A. D., Jr. Former Medical Missionary, Nicaragua.

Wongsarojama, Suwan. Chief of Administration, National Malaria Administration Project. Bangkok, Thailand.

24

3

Clinical Aspects of Scorpion Envenomation

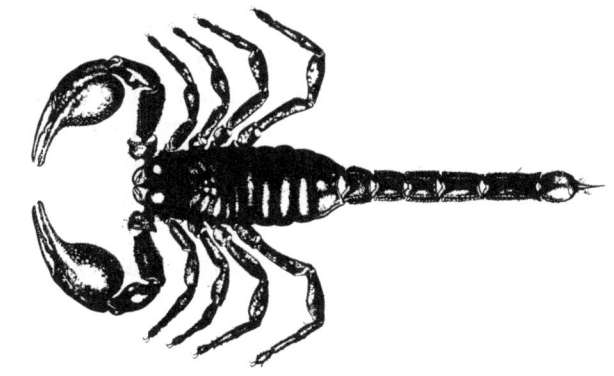

WHILE IMMEDIATE, sharp pain at the site of venom injection is a common feature of all scorpion stings, subsequent developments allow division of these accidents into two categories. In the first of these, symptoms are local and usually transitory, persisting for a few minutes or hours to a day or so. The second category includes those cases showing the systemic involvement characteristic of severe envenomation. With some exceptions, there are no hard and fast guidelines for the separation of scorpion taxa into "harmless" and "dangerous" groups. Some genera, such as *Centruroides* and *Tityus*, may contain dangerously venomous species along with others whose stings produce relatively mild effects. Size and appearance alone are definitely not criteria for judgment of the medical importance of a scorpion. None of the very large and often belligerent scorpions of the Old World genera *Pandinus* and *Heterometrus*, and *Hadrurus* of the New World is regarded as a serious health hazard to man. In contrast, stings by much smaller scorpions of several genera may cause serious illness and even death.

Moderate local edema, which may be discolored, is frequently seen in cases of the milder type of scorpion envenomation. Regional lymph node enlargement, local itching and paresthesia, fever, and occasional nausea and vomiting may occur. Swollen eyelids and "thick tongue," which are sometimes seen, may indicate an allergic response to the venom. In most cases of mild envenomation all signs and symptoms subside within 24 hours. Typical accounts of this type of envenomation have been given by Kopstein (1932), and Stahnke (1956), Marinkelle and Stahnke (1965), Russell *et al* (1968), Rosin (1969), and Williams (1970). Abalos (1963) reported difficulty in swallowing as a symptom following stings by the Argentinian scorpions *Tityus trivittatus* and *T. sectus*. This might represent an intermediate stage of envenomation between the purely local and systemic categories.

Although it is well known that venoms of closely related species of scorpion may differ greatly in toxicity, signs and symptoms reported as following stings by dangerous scorpions in various areas of the world are remarkably similar. Those most frequently encountered are listed in table 1.

While there is general agreement concerning the clinical picture of scorpion envenomation, Balozet (1971) was of the opinion that death from scorpion sting is always due to respiratory paralysis. In contrast, Santhanakrishnan and Balagopal Raju (1974) found that peripheral vascular failure was of importance in this respect. The same authors, as well as Poon-King (1963) and Gueron and Yaron (1970), felt that myocarditis was also an important cause of death in such accidents. The latter were of the opinion that cardiovascular manifestations of envenomation were related to the level of circulating catecholamines elicited by the direct effect of scorpion venom on the sympathetic system. In discussing the convulsions so frequently seen in severe scorpion envenomation, Waterman (1938a and b) wrote that although these had been attributed to direct action of the toxin on the brain, in some cases they were the result of ventricular tachycardia.

Onset of symptoms, other than local pain, may vary from a few minutes to as long as 24 hours following the sting. In their review of 301 cases of stings by *Buthotus tamulus* at Madras, India, Santhanakrishnan and Balagopal Raju (1974) noted that local symptoms were conspicuously absent at the time of admission but that symptoms of peripheral failure were often seen within an hour of scorpion sting. Balozet (1971) noted that the interval between the sting and death varied from several minutes to 30 hours. The confidence interval about the mean was said to be between 2 and 20 hours. He also noted that the physician should be on the alert for the sudden reappearance of symptoms even after the patient seems to be on the road to recovery. Respiratory difficulties are characteristic of such relapses. In reporting 33 deaths in a series of 698 cases of stings by *Tityus trinitatis*, Waterman (1950a) wrote that deaths occurred "at any interval" between 1½ to 42 hours.

While data on pathologic findings in man are rare, Reddy *et al* (1972) reported that disseminated intravascular coagulation and hemorrhages were evident in organs of three adults and four children who had died within 36 hours following stings by the Indian scorpion, *Buthotus tamulus.* Gueron and Yaron (1970) found signs of interstitial edema, frank focal myocarditis, and myocardial necrosis in hearts of fatal cases of envenomation by *Leiurus quinquestriatus.* These patients also showed different degrees of pulmonary edema accompanied by diffuse areas of alveolar hemorrhage.

Mortality from scorpion sting is much higher among children than adults. For example, Waterman (1938 a and b) wrote that the mortality rate due to scorpion sting in Trinidad was 25% among children under five years of age but was only 0.25% for adults. Bücherl (1971) estimated that in Brazil death rates due to stings by *Tityus serrulatus* were 0.8 to 1.4% among adults, 3 to 5% among school children, and 15 to 20% among very young children. Wahbeh (1965) reported that 28 of 87 patients he had treated for scorpion sting in Jordan during 1964 were less than 10 years of age. Five of these children died. He described the scorpion responsible as the yellow color phase of *Buthus occitanus.*

It is generally agreed that administration of a potent antivenin is the most effective treatment of severe envenomation by a scorpion. Various measures have been advocated for use in instances when antivenin is not available, and as adjuncts to serotherapy. Although advice given by Wainschel *et al* (1974) was intended primarily for those involved in treatment of stings by scorpions in the United States of America, their suggestions should also be useful to physicians in any region where scorpions are of public health importance. As Wainschel and his colleagues pointed out, most stings require no special treatment. They advised that a cube of ice over the site of the sting would reduce pain, but that in serious envenomation, both oxygen and positive pressure breathing assistance might be needed. Atropine was recommended for use as a parasympatholytic drug, and it was suggested that when convulsions occur, sodium phenobarbitol given slowly intra-venously is helpful. Stahnke (1956 and 1972) advised that morphine and Demerol not be used for relief of pain in cases of scorpion sting, as these drugs have a synergistic effect with venom of the North American scorpion, *Centruroides sculpturatus,* which greatly increases toxicity of the venom. He did, however, highly recommend the use of barbiturates in treatment. Balozet (1971), writing on treatment of scorpion sting in the Old World, agreed with Stahnke's recommendations concerning morphine and morphine derivatives but states that barbiturates, too, were contraindicated in treatment of scorpion sting because they have an inhibitory effect on the bulbar respiratory centers. Balozet also recommended that patients with severe scorpion envenomation be kept under strict surveillance so that the infrequent sudden relapses may be promptly treated. He noted that in such cases grave respiratory trouble may ensue. Levy (1965) recommended intravenous injections of 20 ml of calcium gluconate for relief of muscle spasms.

In their management of a large series of cases of stings by *Buthus tamulus*, Santhanakrishnan and Balagopal Raju (1974) administered what they called a "lytic cocktail" to all children who presented with peripheral circulatory failure. Prompt and effective treatment was essential, as the venom of this scorpion is highly toxic and there is no specific antivenin available. The "cocktail" used was a freshly prepared solution containing 50 mg of chlorpromazine, 50 mg of promethazine, and 100 mg of pethidine in 50 ml of 5% glucose in distilled water. This was given intravenously in the amount of 0.3 ml per kg/body wt. every 20 minutes after starting maintenance fluids. The therapy was continued until the symptoms of peripheral circulatory failure abated. It was gradually tapered off by half-hourly and hourly intervals. The therapy was usually continued for a maximum of 16 hours.

Steroids (dexamethasone) were also given in the amount of 5 mg every four to six hours to patients who were delayed in admission or to those with associated complications such as myocarditis, pulmonary edema or convulsions. Such therapy was usually given for 24–48 hours. In addition, frusemide was given in the presence of pulmonary edema, digoxin in the event of persistent tachycardia or myocarditis, and antipyretics and sedatives for fever and convulsions.

The authors concluded that the lytic cocktail therapy in combination with steroids was responsible for reduction of the mortality rate from 8.3% several years ago to 2.7% in 1974. They believed that the "deconnection" or semicomotose condition induced by it is capable of mitigating or altering the course of shock by depressing cerebral activity.

Mundle (1961), who treated 78 patients stung by *Buthotus tamulus* during a 14-year period, found that morphine, atropine, glucose saline, glucocorticoids and antibiotics were helpful. He noted that when patients first seen were coughing up copious, blood-tinged, watery, frothy expectoration without exertion, the outcome was invariably fatal.

Although studies by Glenn *et al* (1962) and Potter and Northey (1962) revealed that venoms of several species of scorpion of families Buthidae and Vejovidae possess some antigenic components in common, this information has not yet led to production of a truly polyvalent antivenin which would be effective on a worldwide basis. As McIntosh and Watt (1967) commented, the primary difficulty in preparation of such an antivenin is the poor antigenicity of the neurotoxic components of scorpion venoms. Earlier Whittemore *et al* (1961) had found that most scorpion antivenins are rather narrowly specific in action and tend to neutralize only those venoms which are used in their preparation. Because of this and, perhaps, because production of scorpion antivenin may not be particularly lucrative, antivenins are now produced on a regional basis, primarily under government auspices.

In immunization of animals for preparation of scorpion antivenins, venom solutions of two types are used. In some laboratories, entire telsons are triturated in saline and the resulting filtrate, plus adjuvants, is utilized in the immunizing program. In others, "pure" venom is collected from electrically stimulated scorpions for use in the immunizing process. Therapeutically successful antivenins have been produced with each technique. Manufacturers' instructions for administration vary. Some advise subcutaneous injection; others suggest that either intramuscular or intravenous routes be utilized. Bücherl (1971) urged that antivenin be administered as soon as possible following the sting. He stated that a delay of even two hours could jeopardize success of the treatment.

While Junqua and Vachon (1968) listed 16 institutions where scorpion antivenins were prepared, it is now evident that several of these are no longer in the business. The following list, too, may be out of date at this printing. However, it was compiled through correspondence with directors of institutions where antivenins have been prepared in the past, and at least represents the most recent information available at the time of compilation.

TABLE 1   Characteristics of Severe Scorpion Envenomation in Man

Data on Selected Species of Medical Importance
Envenomation by:

Effects following Stings:	Old World Scorpions Primarily of Africa and the Middle East[1]	Buthotus tamulus[2]	Tityus trinitatis[3]	Centruroides sculpturatus[4]	Leiurus quinquestriatus[5]
Anxiety and/or agitation	X		X	X	X
At site of sting:					
Severe pain	X	X	X	X	
Hyperesthesia				X	
Ecchymosis		X			
Horripilation	X				
Excessive salivation	X		X	X	
Excessive perspiration	X	X	X		X
Vomiting	X		X		
Diarrhea	X				
Melena	X				
Hyper- or Hypotension	X				X
Cold, clammy skin		X	X		
Irregular pulse	X	X	X		
Unstable temperature	X	X		X	
Involuntary micturation and/or defecation	X			X	
Pulmonary edema		X			X
Respiratory difficulties ranging from irregular movements to respiratory paralysis	X	X		X	X
Muscle tone increases ranging from twitching and contractures in abdomen, limbs, and pharynx to convulsions.	X	X	X	X	

[1] Balozet (1971). Deals primarily with species of Androctonus, *Buthus, Buthotus,* and *Leiurus.*

[2] Mundle (1961); and Santhanakrishnan and Balagopal Raju (1974). *Buthotus tamulus* is probably the most dangerous scorpion of India.

[3] Waterman (1938a and b, 1950a and b); Poon-King (1963). *Tityus trinitatis* is a species of considerable medical importance in Trinidad.

[4] Stahnke (1972). *Centruroides sculpturatus* is regarded as the dangerous scorpion of the United States. It is found in northern Mexico and in Arizona and possibly portions of adjoining states.

[5] Gueron and Yaron (1970). *Leiurus quinquestriatus* is one of the most dangerous scorpions of North Africa and the Middle East. Perhaps because these authors had the "broad picture" in mind, they did not mention some of the individual signs and symptoms which are associated with hypertension, peripheral vascular collapse, congestive heart failure, and pulmonary edema.

	TABLE 1 (cont.)				
Effects following stings:	Old World Scorpions, Primarily of Africa and the Middle East	Buthotus tamulus	Tityus trinitatis	Centruroides sculpturatus	Leiurus quinquestriatus
Hematuria	X				
Hyperglycemia	X		X		X
Glycosuria	X		X		
Myocarditis		X	X		X
VMA elevation					X
SGOT elevation					X
Priapism	X			X	
Exophthalmia	X				
Syncope, delirium, mental cloudiness	X				
Hemiplegia			X		
Shock		X	X		X
Blurred vision, mydriasis, blindness	X		X	X	
Oliguria	X				
Polyuria	X				

CURRENT ANTIVENIN PRODUCTION

1. Antivenom Production Laboratory, Arizona State University, Tempe, Arizona 85281. The scorpion antivenin produced here is effective in treatment of stings by *Centruroides sculpturatus*, the only dangerously venomous scorpion in the United States of America. The antivenin is lyophilized and is prepared through immunization of goats with venom of *C. sculpturatus*. This antivenin is distributed free throughout the state of Arizona.

Note: In a brochure of this laboratory published in 1973, it was advised that supportive therapy before administration of antivenin should include barbiturates and atropine.

2. Laboratorios "Myn," S. A., Avenida Coyoacan No. 1707, Mexico 12, D. F. Two scorpion antivenins, one lyophilized, the other in liquid form, are distributed by this company. Both products are produced immunizing horses with venom solutions resulting from trituration of whole telsons of Mexican species of genus *Centruroides*. While they will also neutralize venoms of species of *Centruroides* found in Central America and that of *C. sculpturatus*, which occurs in Mexico as well as the southwestern United States, a special

permit from the Center for Disease Control in Atlanta is required for importation of such items into the United States.

3. Instituto Butantan, Caixa Postal 65, Sao Paulo, Brazil. The bivalent scorpion antivenin produced at this institution is effective against venoms of two most dangerous South American scorpion species, *Tityus serrulatus* and *T. bahiensis*. It may also neutralize venoms of other species of genus *Tityus* in Central and South America but data on this subject are not available. The antivenin, in liquid form, is produced by immunizing horses.

4. Reyfik Saydan, Central Institute of Hygiene, Ankara, Turkey. The antivenin produced here is effective against venom of *Androctonus crassicauda*, the most dangerous scorpion in Turkey. Whittemore *et al* (1961) found that in neutralization tests using white mice, this product was also effective against venoms of the Algerian species, *Androctonus australis*, North African, Southern European *Buthus occitanus*, and the South American *Tityus serrulatus* and *T. bahiensis*.

5. Institut Pasteur d'Algérie, Rue du Dr. Laveran, Algiers, Algeria. The antivenin prepared here is effective in treatment of stings by *Androctonus australis* and *Buthus occitanus*. It is (or was) supplied in 10 ml ampules in liquid form.

6. State Serum and Vaccine Institute, Agouza, Cairo, Egypt. The monovalent antivenin prepared here is effective in treatment of stings by *Leiurus quinquestriatus*, the most dangerous scorpion in Egypt.

7. South African Institute for Medical Research, Johannesburg, P. O. Box 1038, South Africa. Only species of genus *Parabuthus* are of medical importance in South Africa, and even they are much less dangerous than scorpions of genera *Leiurus*, *Androctonus*, and *Buthus* of North Africa and the Middle East. The antivenin prepared here is in liquid form.

References on Clinical Aspects of Scorpion Envenomation

Abalos. J. W. 1963. Scorpions of Argentina. *In: Venomous and Poisonous Animals and Noxious Plants of the Pacific Region*. Keegan, H. L. and MacFarlane, W. V., eds. Pergamon Press, Oxford, pp. 111–117.

Balozet, L. 1971. Scorpionism in the Old World. *In: Venomous Animals and Their Venoms*, vol. III, *Venomous Invertebrates*. Bücherl, W. and Buckley, E. E., eds. Academic Press, New York, London, pp. 349–371.

Bücherl, W. 1971. Classification, Biology, and Venom Extraction of Scorpions. Ch. 55 *In: Venomous Animals and Their Venoms*, vol III, *Venomous Invertebrates*. Bücherl, W. and Buckley, E. E., eds. Academic Press, New York, London, pp. 317–347.

Glenn, W. G., Keegan, H. L., and Whittemore, F. W., Jr. 1962. Intergeneric relationships among various scorpion venoms and antivenins. *Science, 135:* 433–435.

Gueron, M. and Yaron, R. 1970. Cardiovascular manifestations of severe scorpion sting. *Chest, 57:* 156–162.

Junqua, C. and Vachon, M. 1968. *Les Arachnides Venimeux et Leurs Venins.* État actuel des recherches. Académie royale des Sciences d'Outre-Mer. Classe des Sciences naturelles et médicales, N. S. XVII-5. Brussells, 136 pp.

Kopstein, F. 1932. Die giftiere Java's und ihre bedeutung fur den Menschen. *Mededeel. v. d. dienst. f. volksgezondh. in Nederl. Indiie, 21:* 222–256.

Levy, M. 1965. Les envenimations par piqûres de scorpions à Marrakech. (Maroc). Thèse, Faculté de Médecine, Paris, 42 pp.

Marinkelle, C. J. and Stahnke, H. L. 1965. Toxicological and clinical studies on *Centruroides margaritatus* (Gervais), a common scorpion in western Colombia. *J. Med. Ent., 2:* 197–199.

McIntosh, M. E. and Watt, D. D. 1967. Biochemical-immunochemical aspects of the venom from the scorpion *Centruroides sculpturatus*. In: Animal Toxins. Russell, F. E. and Saunders, P. R., Eds. Pergamon Press, Oxford, New York, pp. 47–58.

Mundle, P. M. 1961. Scorpion stings. *Brit. Med. J., 1:* 1042.

Poon-King, T. 1963. Myocarditis from scorpion stings. *Brit. Med J., 1:* 374–377.

Potter, J. M. and Northey, W. T. 1962. An immunological evaluation of scorpion venoms. *Am. J. of Trop. Med. and Hyg., 2:* 712–716.

Reddy, C. R. R. M., Suvarnakum, G., Devi, C. C., and Reddy, C. N. 1972. Pathology of scorpion venom poisoning. *J. Trop. Med. and Hyg., 75:* 98–100.

Rosin, R. 1969. Sting of the scorpion *Nebo hierichonticus* in man. *Toxicon, 7:* 75.

Russell, F. E., Alender, C. B. and Buess, F. W. 1968. Venom of the scorpion *Vejovis spinigerus. Science, 159:* 90–91.

Santhanakrishnan, B. R. and Balagopal Raju, V. 1974. Management of scorpion sting in children. *J. Trop. Med. Hyg., 77:* 133–135.

Stahnke, H. L. 1956. *Scorpions.* 2nd Edition. Poisonous Animals Research Laboratory. Arizona State College, Tempe, 36 pp.

———. 1972. Arizona's lethal scorpion. *Ariz. Med., 29:* 490–493.

Wainschel, J., Russell, F. E. and Gertsch, W. S. 1974. Bites of spiders and other arthropods. *Current Therapy.* H. F. Conn, ed. W. B. Saunders Company, pp. 865–867.

Waterman, J. A. 1938a. Some notes on scorpion poisoning in Trinidad. *Trans. Roy. Soc. Trop. Med. & Hyg., 31:* 607–624.

———. 1938b. A few cases. *Caribbean Med. J., 1:* 119–120.

———. 1950a. Two cases of scorpion poisoning characterized by convulsions with electrocardiograms. *Caribbean Med. J., XII:* 127–129.

———. 1950b. Scorpions in the West Indies with special reference to *Tityus trinitatis. Caribbean Med. J., XII:* 167–177.

Wahbeh, Y. 1965. Scorpion stings in children. *Jordan Med., J., 1:* 57–61.

Whittemore, F. W., Jr., Keegan, H. L., and Borowitz, J. L. 1961. Studies of scorpion antivenins. 1. Paraspecificity. *Bull. Wld. Hlth. Org., 25:* 185–188.

Williams, S. C. 1970. The effects on man of a natural sting by the scorpion *Vejovis confusus* Stahnke. *Pan-Pacific Entomol., 46:* 77–78.

34

4

Scorpion Control and Prevention of Scorpion Stings

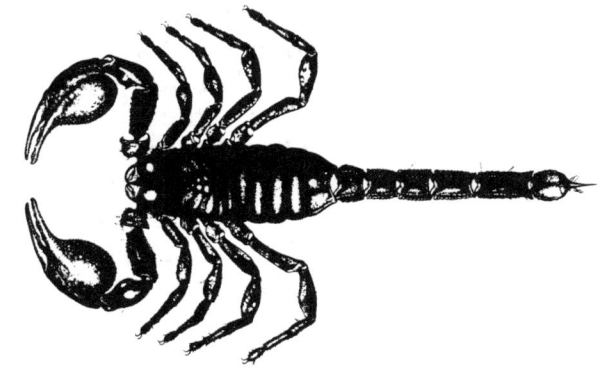

EFFORTS TO ELIMINATE scorpion sting as a medical problem have included: destruction of scorpions with pesticides; mechanical barriers to prevent entry of scorpions into buildings; general "policing" of yards around homes to remove litter which might harbor scorpions; use (primarily experimental) of predators; and common sense measures to be followed by campers or others who might be living outdoors in areas where scorpions occur.

Much of the earlier work on effects of insecticides on scorpions involved the chlorinated hydrocarbons. Thompson (1946) was the first to test effectiveness of DDT against these arachnids. Toxicity of related compounds for species of *Tityus* was tested by de Souza *et al* (1954) and Bücherl (1956). Ridgway *et al* (1961) and McGregor and Flanigan (1962) found lindane more effective than either chlordane or malathion against *Centruroides vittatus*, a relatively innocuous species from Texas. More recently, Rolli (1972) tested lindane, along with other insecticides, in an effort to find a less persistent substitute for BHC, which was then used for scorpion control in Tunisia. All insecticides were tested in powder formulations. Twenty-four hours following a three-minute exposure period on treated sand, it was found that both 25% BHC and 1% lindane had produced 100% kills of the dangerously venomous *Buthus occitanus* and *Androctonus crassicauda*. A 1% parathion powder also proved highly lethal for *B. occitanus*. The author claimed satisfactory results in field tests with 1% lindane.

As a result of policies established by the Environmental Protection Agency in June 1972, and passage of the Federal Environmental Pesticide Control Act in October 1972, use of some insecticides, such as DDT, has been almost entirely prohibited in the United States, and restrictions have been placed upon the sale and availability of other insecticides to the public. However, these regulations, which were designed to reduce hazards to man and other nontarget organisms in the environment, have not greatly affected availability of pesticides for scorpion control. Other information on pesticides is given in two books—*Pesticide Handbook–Entoma* and *Pest Control.* Both are commercially sponsored.

In 1973, insecticide formulations recommended for indoor use for scorpion control included residual sprays containing 1-2% carbaryl, 2% chlordane, 1.5% dieldrin, or 5% deodorized malathion, or dusts containing 2% carbaryl, 5% chlordane, or 1% dieldrin. It was recommended that the sprays be used for spot treatment of baseboards, under furniture, closets, crevices around plumbing, and any areas where scorpions are seen. Attics and crawlways, as well as basements, should also be treated. Dusts were said to be particularly useful in wall voids. The long residual life of chlordane and dieldrin makes them materials of choice for this purpose. It was mentioned that either 1% dusts or 0.5% sprays of lindane tend to excite scorpions and thus might increase the danger of stings until the insecticides take effect. Recommendations for outdoor applications included sprays containing 2% carbaryl, 2% chlordane, and 0.5% dieldrin, or dusts containing 10%, 10% and 2%, respectively, of these insecticides. Granules of diazonon (10%) were also recommended for area treatment. Outdoor treatment should include structures such as out-buildings with surfaces in contact with the soil. In such cases, residual sprays should be applied to a height of 2 feet above ground level. Hiding places such as stacks of lumber, debris, and firewood should also be treated.

While the author has no reason for doubting the value of these recommendations, it should be noted that they were given without reference to results of either laboratory or field tests of their effectiveness. The 1974 release of "Public Health Pesticides," published by the Center for Disease Control of the Public Health Service, included recommendations for the same pesticides listed above, but did not mention formulations.

Observations by Mazzotti (1962a, 1962b, and 1964b) that the Mexican scorpions, *Centruroides limpidus* and *C. suffusus*, were unable to climb up smooth surfaces such as glazed ceramic mosaic, led him to make some specific recommendations for prevention of entry of scorpions into houses. There were three basic recommendations: 1. The level of the floor of the house must be above the ground in order to have a threshold on every door at least 20 cm in height, forming one step. 2. The front of each step and the entire perimeter of the house should be protected by a continuous horizontal row of glazed ceramic mosaic. 3. It is advisable not to grow plants which touch the walls of the house and offer pathways for scorpions into the home. In addition to the protection with glazed ceramic tile, it was recommended that the exterior walls of buildings be well smoothed with planed cement up to a height of 60 cm. If tile is not available or is too expensive, a continuous horizontal strip of aluminum foil about 5 cm in height should be placed on the outer and inner walls of the house just below the roof as an emergency measure. While Mazzotti's recommendations were designed for application in Mexico where scorpions hide in the palm, straw, or tile roofs of houses, his idea could well be of practical importance in other regions where construction is similar and scorpion stings are numerous. Regardless of the type of mechanical barrier installed, residual insecticide applications should also be made. Effectiveness of this combined approach in reduction in numbers of scorpion stings in Coahuayana, Michoacan, Mexico was demonstrated by Bravo-Becherelle and Arizmendi (1967).

In addition to the specific measures listed above, it is important that potential scorpion harborages such as piles of lumber or bricks and other debris be removed from the immediate vicinity of houses and periodically treated with insecticides. Persons camping out in areas where scorpions occur should be sure to shake out shoes or other items of clothing or equipment which may have been lying on the ground overnight. The author recalls one incident at Brownsville, Texas, in which 26 specimens of *Centruroides vittatus* were found among the folds of one small tent which had been packed at the bivouac area during a shower and later unrolled to dry.

Although Mazzotti (1966) found that chickens, ducks, and cats will eat scorpions and Ennik (1972) listed a variety of wild mammals, birds, reptiles, and even insects (Jerusalem crickets and darkling beetles) which will prey upon scorpions, there is no published evidence that natural predators have been truly effective in control of these arachnids. Mazzotti (1964a) suggested that the tarantula, *Aphonopelma smithi*, might be an effective predator against scorpions in some situations in Mexico. However, in laboratory experiments by Wheeling and Keegan (1971) using the Mexican scorpion, *Centruroides limpidus tecomanus*, it was found that while most tarantulas would attack scorpions, one sting was sufficient to make them lose all interest in feeding. It was of interest though that tarantulas that had been stung were sluggish for a few hours but suffered no lasting effects. The same results occurred when tarantulas were given injections of venom in amounts that had proven lethal for mice weighing 30g. It seems highly probable that even if tarantulas proved to be efficient predators, this method of control would hardly have caught on. The author suspects that to most people there would seem to be little choice between the scorpion and the tarantula, even though the latter is harmless to man.

References on
Scorpion Control
and Prevention of
Scorpion Stings

Anonymous. 1973. Public health pesticides. *Pest Control, 41:* 17–50. This publication was prepared at the Technical Development Laboratories, Center for Disease Control, Public Health Service, Savannah, Georgia 31402.

———. 1974. Public health pesticides. This appeared in the June 1974 issue of *Pest Control*, pp. 30–31. Other publication data as indicated above except that this material is now prepared at the Bureau of Tropical Diseases, Center for Disease Control, Atlanta, Georgia 30333.

Bravo-Becherelle, M. A. and Arizmendi, N. 1967. Valoracion de la proteccion mecanica de las casas contr la entrada de alacranes. *Salud Publica Mexico, 9:* 209–211.

Bücherl, W. 1956. Scorpions and their effects in Brazil. IV. Observations on insecticides lethal to scorpions and other methods of control. *Mem. Inst. Butantan, 27:* 107–120. (In Portuguese).

Ennick, F. 1972. A short review of scorpion biology, management of stings and control. *California Vector Views, 19:* 69–80.

Mazzotti, L. 1962a. Proteccion mecanica de las casas contra la entrada de alacranes (escorpiones). *Rev. Inst. Salubr. Enferm. trop., XXII:* 183–198.

———. 1962b. Procedimiento para investigar el grado de apitud que tienen algunos arthropodos par trepar por diversos de superficies. *Rev. Inst. Salubr. Enferm. trop., XXII:* 199–201

———. 1964a. Enemigoes de los alacranes: Tarantula del genero *Aphonopelma. Rev. Inst. Salubr. Enferm. trop., XXIV:* 9–10.

————. 1964b. Medidas conplementarias en relacion con la proteccion mecanica de los edificios contra los alacranes. *Rev. Inst. Salubr, Enferm. trop., XXIV:* 11–14.

————. 1966. Estudio sobre enemigos naturales de los alacranes. *Rev. Invest. Salud Publica, 26:* 51–55.

McGregor, T. and Flanigan, J. F. 1962. Effects of insecticides on the scorpion *Centruroides vittatus. J. Econ. Ent., 55:* 661–662.

Ridgway, R. L., McGregor, T. and Flanigan, J. F. 1961. The effect of residual deposits of insecticides on the scorpion, *Centruroides vittatus. J. Econ. Ent., 55:* 1012–1013.

Rolli, K. 1972. Essais de differents insecticides dans la destruction des scorpions. *Arch. de l'Institute Pasteur de Tunis, 49:* 267–274.

Souza, J. C. de, Bustamente, F. M. de, and Bicalho, J. C. 1954. Novos dados sobre a combates aos escorpioes em Belo Horizonte com o hexochlorociclohexana. *Rev. Brasil Malariol. Doencas Trop., 6:* 357–361.

Thompson, G. A. 1946. Effect of DDT on scorpions. *J. Econ. Ent., 39:* 357.

Wheeling, C. H. and Keegan, H. L. 1971. Effects of a scorpion venom on a tarantula. *Toxicon, 10:* 305–306.

5

Classification of Scorpions

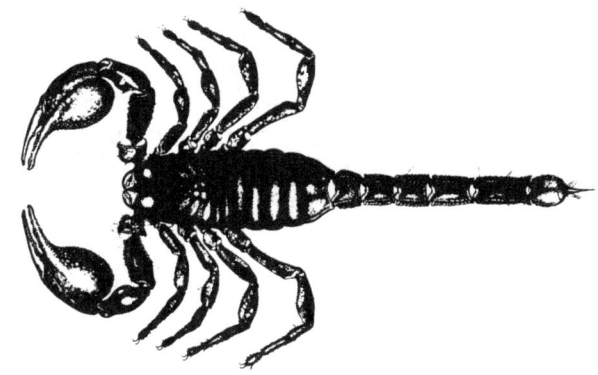

WHILE AUTHORITIES have agreed that either five or six families of the order Scorpionida should be recognized, there has been some difference of opinion concerning the status of two or three of these. At present, the consensus seems to be that the following should be considered as "fullfledged" families: Buthidae, Scorpionidae, Diplocentridae, Chactidae, Vejovidae and Bothriuridae. Characteristics of each of these follow:

1. BUTHIDAE

Sternum triangular; an accessory spine usually present on telson; both external and internal pedal spurs present; tibial spurs usually present, but may be vestigial or absent in some species: three to five lateral eyes on either side of the carapace.

Genera of this family include the most dangerous species of the order. They are found in both the Old and New Worlds. Examples are: *Buthus, Androctonus, Leiurus, Buthotus, Centruroides, Tityus, Parabuthus,* and *Mesobuthus.*

2. SCORPIONIDAE

Sternum pentagonal; no accessory spine on telson; external pedal spurs only; tibial spurs lacking on leg pairs III and IV; terminal margins of lateral lobes of tarsomere II rounded in some species.

With the exception of *Hemiscorpion lepturus*, a Middle Eastern species, it appears that none of the members of this predominantly Old World family is dangerously venomous for man. Species of some genera are greatly feared because of their size and their belligerent posture when disturbed. Foremost in this category are species of the African genus *Pandinus* and the Central and Southeast Asian *Heterometrus.*

41

3. DIPLOCENTRIDAE

Differs from the Scorpionidae mainly in that there is always an accessory spine on the telson.

The sole representative of the family which has been mentioned as of even minor medical importance is the Middle Eastern *Nebo hierochonticus*. Three species of genus *Diplocentrus* occur in the United States, one in Florida, the others in the Southwest.

4. CHACTIDAE

Sternum pentagonal; no accessory spine on telson; both external and internal pedal spurs present; tibial spurs lacking; two lateral eyes on either side of the cephalothorax, rarely eyeless; spiracles oval.

None of the species of this family is of significant medical importance.

5. VEJOVIDAE

Sternum pentagonal, usually broader than long, and with a median depression; no accessory spine on telson; both external and internal pedal spurs present; no tibial spurs; three lateral eyes on either side of the carapace; spiracles are frequently oval.

Most species of this family occur in the New World, particularly in the western United States, Mexico, Central America, and portions of South America. While scorpions of several species of the genera *Vejovis* and *Hadrurus* can deliver painful stings, the effect is nearly always local.

6. BOTHRIURIDAE

Members of this family are unique in that the sternum consists of two transverse bars, and is several times broader than long. Bücherl (1971) wrote that in some species the sternum is scarcely visible. Although many species of genus *Bothriurus* occur in South America, particularly in Brazil, Argentina and Chile, there is no published information to indicate that any of these is of other than minor medical importance.

6

Accounts of Genera and Species

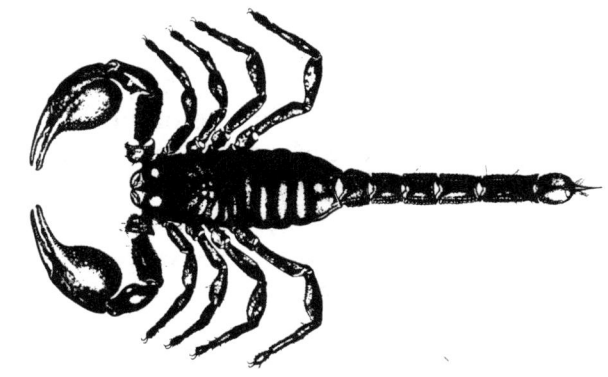

IT IS PROBABLE that approximately 800 scorpion species have been described and that less than 50 of these can be considered as dangerously venomous. It is possible, of course, that many species of scorpion whose habits do not bring them into contact with man may possess highly potent toxins. However, an estimation of the public health importance of a scorpion species, in the opinion of the writer, should not be based solely on the degree of severity of envenomation resulting from its sting, but also on the fact that arachnids, and particularly spiders and scorpions, rank high among the arthropods as stimuli for entomophobia. This situation, created by folklore, dramatic but inaccurate literature, and TV and motion picture sequences, may actually be of value in areas where dangerously venomous species exist. However, a less desirable effect has been that much needless apprehension has been created in regions where only relatively innocuous scorpions come into contact with man. Merely because of their size, scorpions of several genera and species which commonly occur in association with man are greatly feared. For this reason, illustrations of some of these are included in this publication. In addition, information is given concerning their morphology, habits, geographic distribution and, if known, effects of their sting on man.

FAMILY BUTHIDAE

Genus *Androctonus*
(Hemprich and
Ehrenberg), 1829

Scorpions of genus *Androctonus*, commonly known as the "fat tailed" scorpions, are so named because of the marked thickness and width of the segments of the postabdomen. In addition, a frequently deep dorsal concavity on each postabdominal segment contributes to the extreme development of the dorsal keels. The latter are often serrated. Among the eight species and numerous subspecies which have been described are several scorpions of considerable medical importance. Species of the genus occur in Northwest Africa and in the Middle East from Turkey south and westward to the borders of India and Pakistan. Examples of only three species of the genus were available for study and illustration. Distribution of species of the genus has been given by Vachon (1966).

Androctonus australis (Linneus), 1758 [PLATE VI]

This formidable scorpion not only possesses the most potent venom of the species of *Androctonus* but, unfortunately, is also the most widely distributed member of the genus. *Androctonus australis* and its subspecies are found in North Africa in Egypt, Algeria, Tunisia and Libya. While the species has not been reported from Iran and other countries of the Middle East, this may be due to a lack of collecting effort as Pocock (1900) described two subspecies of *australis* from Pakistan and eastern India. The color of the adults varies from brown to yellow in different geographic areas. The pedipalps, the fifth, and sometimes the fourth segment of the postabdomen, are darker than the remainder of the animal. Minning and Zumpt (1942) wrote that the species was found most frequently in the arid mountainous regions of North Africa and seemed to avoid the more moist coastal areas. This was borne out by a list of specific collection localities given by Vachon (1952). Many of these are on high plateaus of the Atlas Mountains in Algeria.

Although drop-for-drop the venom of *A. australis* is less potent than that of *Leiurus quinquestriatus*, at least for white mice, the greater venom yield of the larger *australis* renders the two species about on a par with respect to danger to man. In some areas, such as southern Algeria, the great majority of scorpion stings are due to *A. australis*. Balozet (1964), citing records of the Service de Sante des Territoires du Sud (Algeria), reported 20,164 cases of scorpion sting, with 386 deaths during the period 1942–1958. The death rate was conspicuously higher among infants and small children than among adults. Balozet (1971) found that the toxicity of venom of *A. australis* varied with the geographic origin of the scorpions. Effects of envenomation by *australis* on man are described in chapter 3, "Clinical Aspects of Scorpion Envenomation," in this publication.

An antivenin for treatment of stings by *A. australis* and *Buthus occitanus* was produced for many years at the Institut Pasteur in Algiers, however, there is no current information concerning the present status of its production. Whittemore, Keegan, and Borowitz (1961) found an antivenin produced with venom of *Androctonus crassicauda* in Turkey more effective than a homologous antivenin from Algiers in protecting white mice against effects of *australis* envenomation.

Adults of *A. australis* often reach a length of 10 cm or more. The specimen figured in plate 6 was 96 mm long. This specimen, along with several others, was obtained in 1959 through the kindness of Dr. Lucien Balozet, of the Institut Pasteur d/Algerie. It is assumed that the specimen figured was collected in Algeria.

Androctonus crassicauda (Olivier), 1807 [PLATE VII]

The geographic range of this scorpion is similar to that of *Leiurus quinquestriatus* in much of the Middle East and North Africa. *Androctonus crassicauda* is found from Turkey southward in the Middle East, where it has been recorded from Syria, Jordan, Iraq, Iran, Israel, and Arabia. Balozet (1971) mentions its occurrence in Egypt and Sudan, while Vachon (1952) described a subspecies, *A. c. gonneti* found only in Morocco. Its color ranges from dark brown to black except that the terminal segments of the legs and fingers of the pedipalps may be greenish or almost yellow in some cases. Vachon (1952) noted that specimens of *crassicauda* in the Middle East were generally lighter in color than those in Morocco. As in *A. australis*, the dorsal concavity on caudal segments two, three and four is particularly deep, more so than in most other species of the genus. The species also resembles *australis* in that the median lateral keels of segments two and three of the postabdomen are reduced to only a few granules.

Tulga (1960 and 1964) noted that while *A. crassicauda* was the most dangerous scorpion in most of Turkey, venom of this species was only about 20% as potent as that of *Leiurus quinquestriatus* as determined in laboratory tests using white rats. He also wrote that although *A. crassicauda* antivenin neutralized both venom of the homologous species and that of *L. quinquestriatus*, antivenin prepared using venom of the latter species did not neutralize *crassicauda* venom. Whittemore, Keegan, and Borowitz (1961) found that an antivenin produced with venom of *A. crassicauda* at the Central Institute of Hygiene, Ankara, Turkey was more effective than

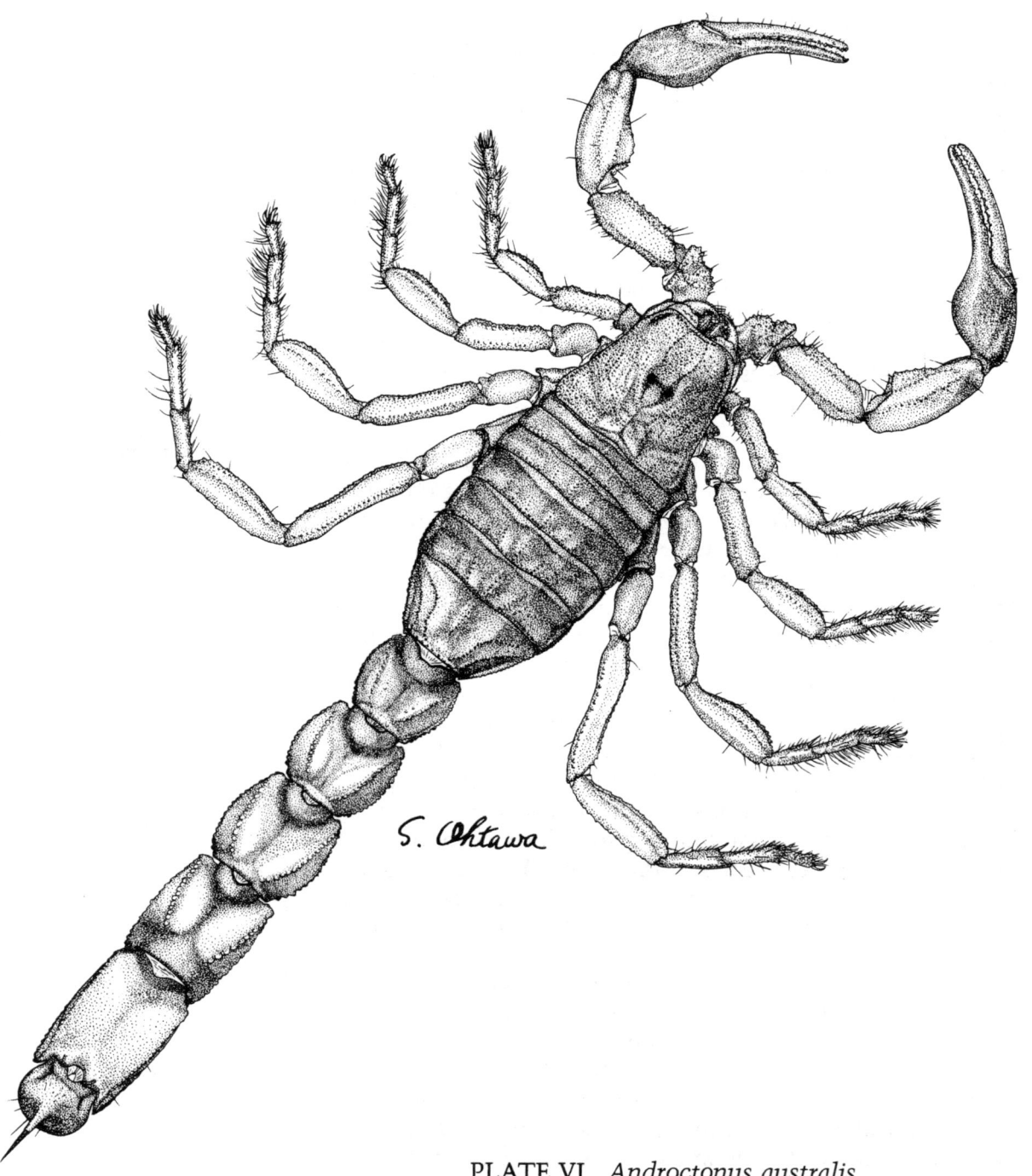

PLATE VI *Androctonus australis*
(Linneus), 1758

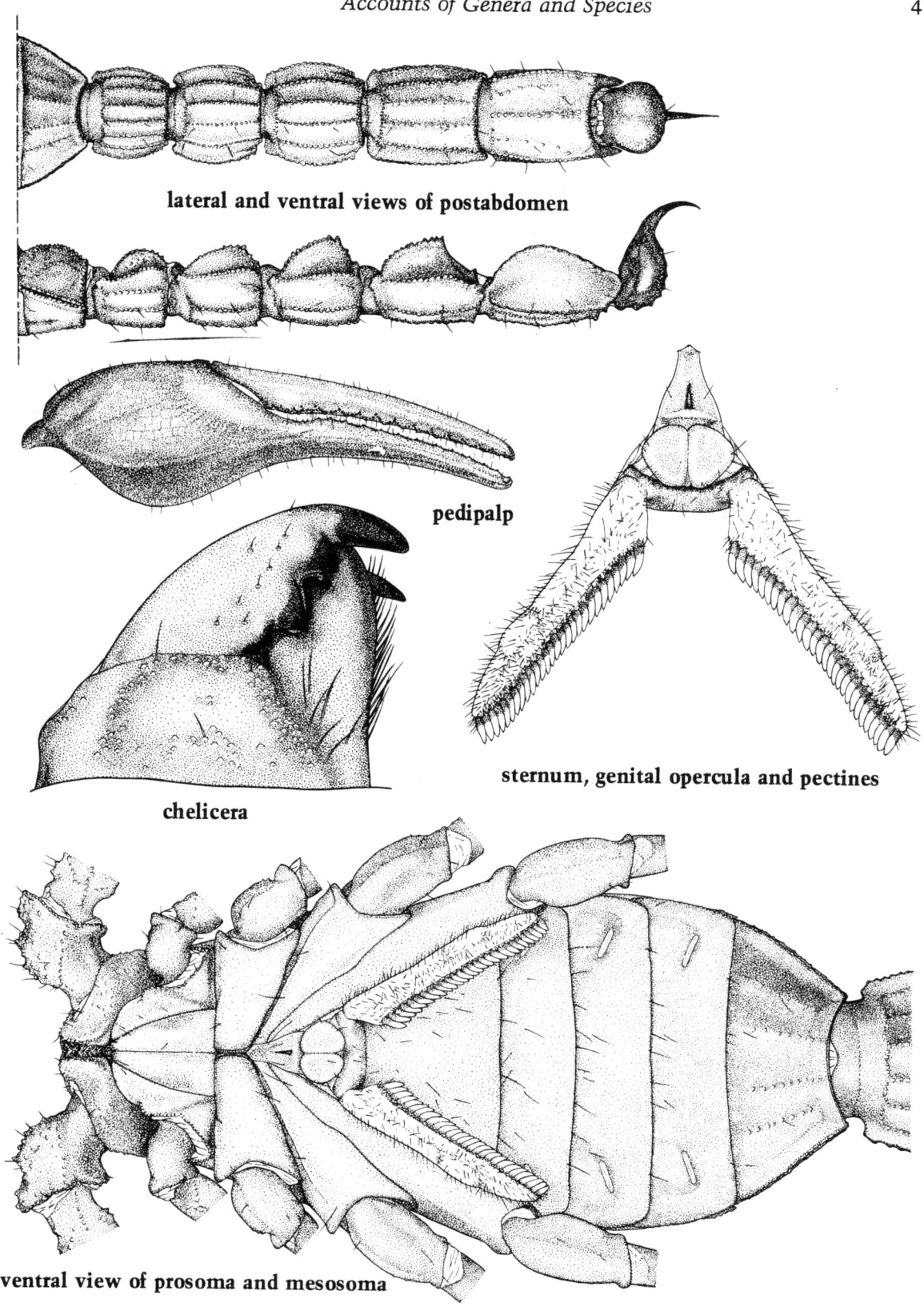

lateral and ventral views of postabdomen

pedipalp

chelicera

sternum, genital opercula and pectines

ventral view of prosoma and mesosoma

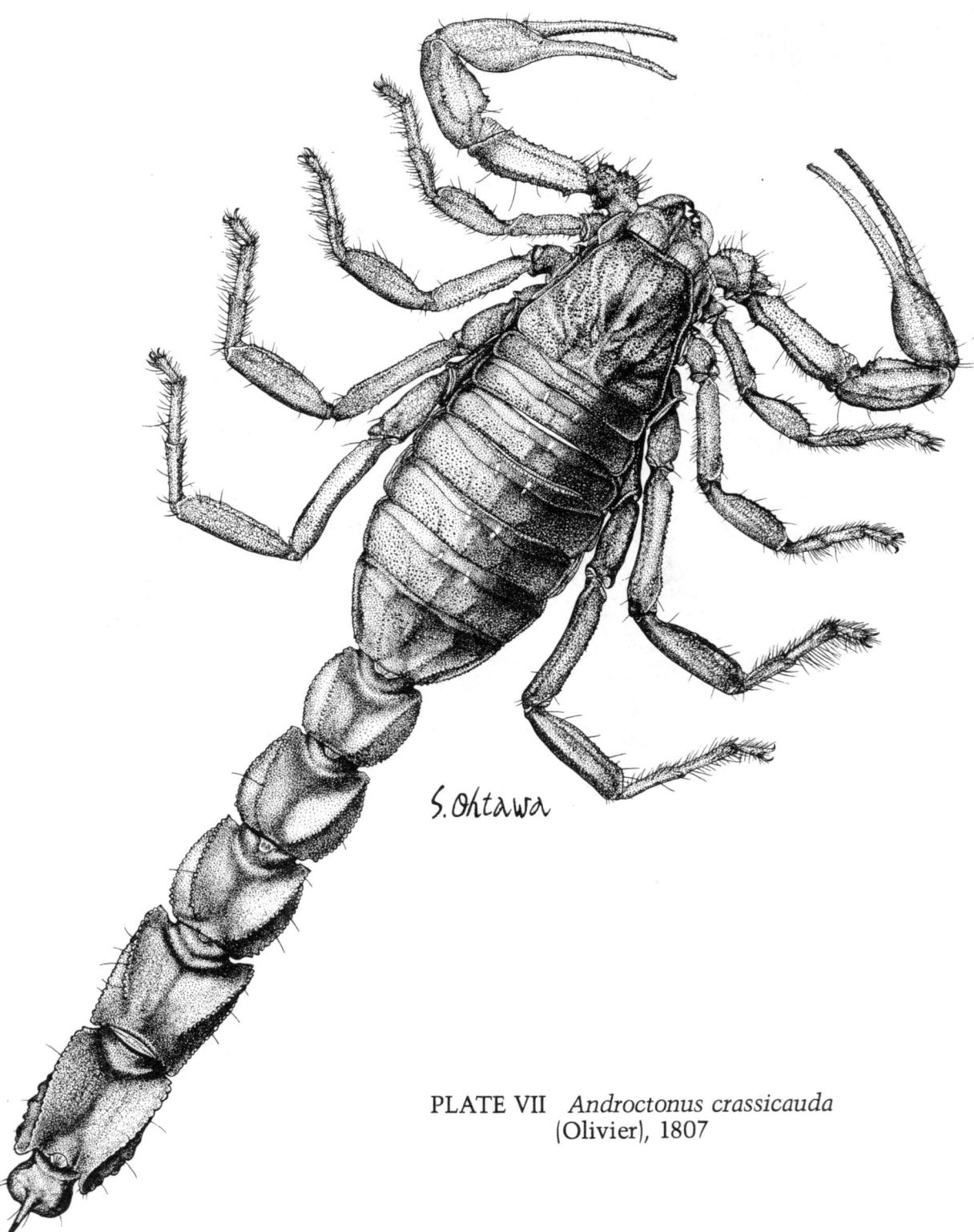

S. Ohtawa

PLATE VII *Androctonus crassicauda*
(Olivier), 1807

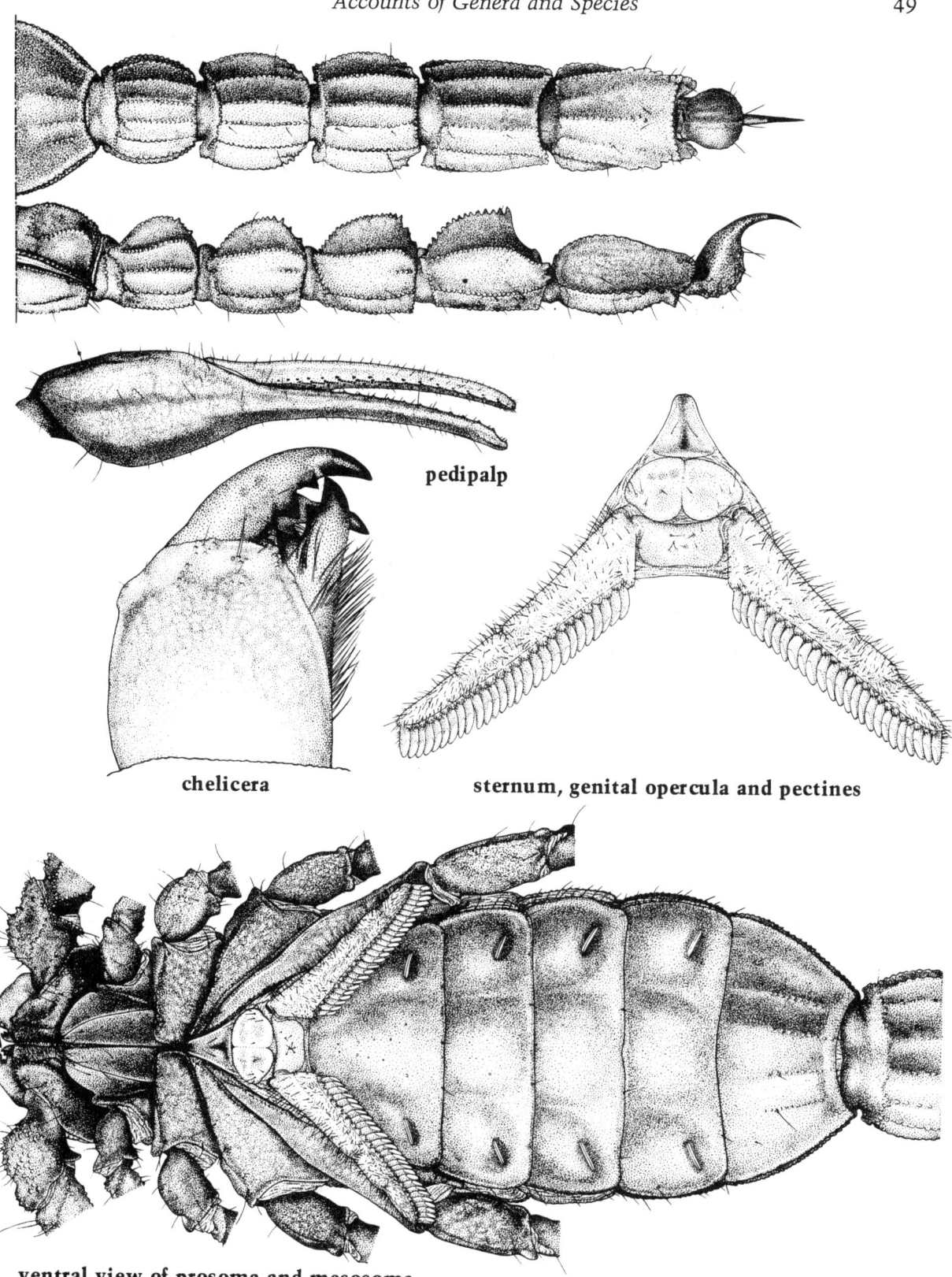

pedipalp

chelicera

sternum, genital opercula and pectines

ventral view of prosoma and mesosoma

homologous antivenins in neutralization of venoms of *Buthus occitanus* and the South American species, *Tityus serrulatus* and *T. bahiensis*. Such paraspecificity is rare.

Adults of *A. crassicauda* species are frequently over 5 cm in length. The specimen figured in plate 7 was 73 mm in length. This specimen, as well as venom and antivenin for study, was supplied through the courtesy of Dr. Turgut Tulga of the Central Institute of Hygiene, Ankara, Turkey in 1959.

Androctonus bicolor (Hemprich and Ehrenberg), 1829 [PLATE VIII, AND FIGURE 2]

Androctonus bicolor occurs, along with *A. crassicauda*, in Egypt, Israel, and Jordan. Shulov (1962) wrote that *A. bicolor* was found in all regions of Israel, but was more abundant in the south. The color, like that of *A. crassicauda*, is dark, except that the terminal segments of the legs and pedipalps are lighter. It differs morphologically from *crassicauda* and from other species of the genus in that the hand of the pedipalp is scarcely wider than the closed fingers, and not as wide as the patella. In specimens examined by the writer, it was also found to be unlike *crassicauda* in that the median lateral keels of postabdominal segments two and three are almost completely developed instead of being restricted to only a few granules.

Although Shulov (1952) noted that *A. bicolor*, along with *A. crassicauda*, possessed "quite strong venom," he had earlier (1939) written that of the scorpions of Israel, only *L. quinquestriatus* could be considered as dangerous. No information is available concerning effects of stings by *A. bicolor* on man or on the effectiveness of heterologous antivenins in neutralizing venom of this species.

In Israel, the species is found under rocks, mainly in hilly, arid regions. In March 1962, the writer, along with Professor Aharon Shulov of the Hebrew University of Jerusalem and several of his colleagues, collected numerous specimens of *bicolor* about 40 km south of Beersheba. The species is reported to attain a length of 80 mm. The specimen figured on plate 8 was 72 mm in length.

Other Species of the Genus

The five species of genus *Androctonus* not illustrated in this publication are: *A. aeneas*, C. L. Koch, 1839, found in Tunisia, Morocco and Algeria; *A. amoreuxi* (Audouin and Savigny), 1812 and 1826 found in the Middle East in Iran and Israel, and in North Africa in Egypt, Morocco, Algeria, Chad, Sudan, Senegal, and Tunisia; *A. hoggarensis* (Pallary), 1929 in Algeria and Nigeria; *A. mauretanicus* (Pocock), 1902 and its two sub-

FIGURE 1

FIGURE 2

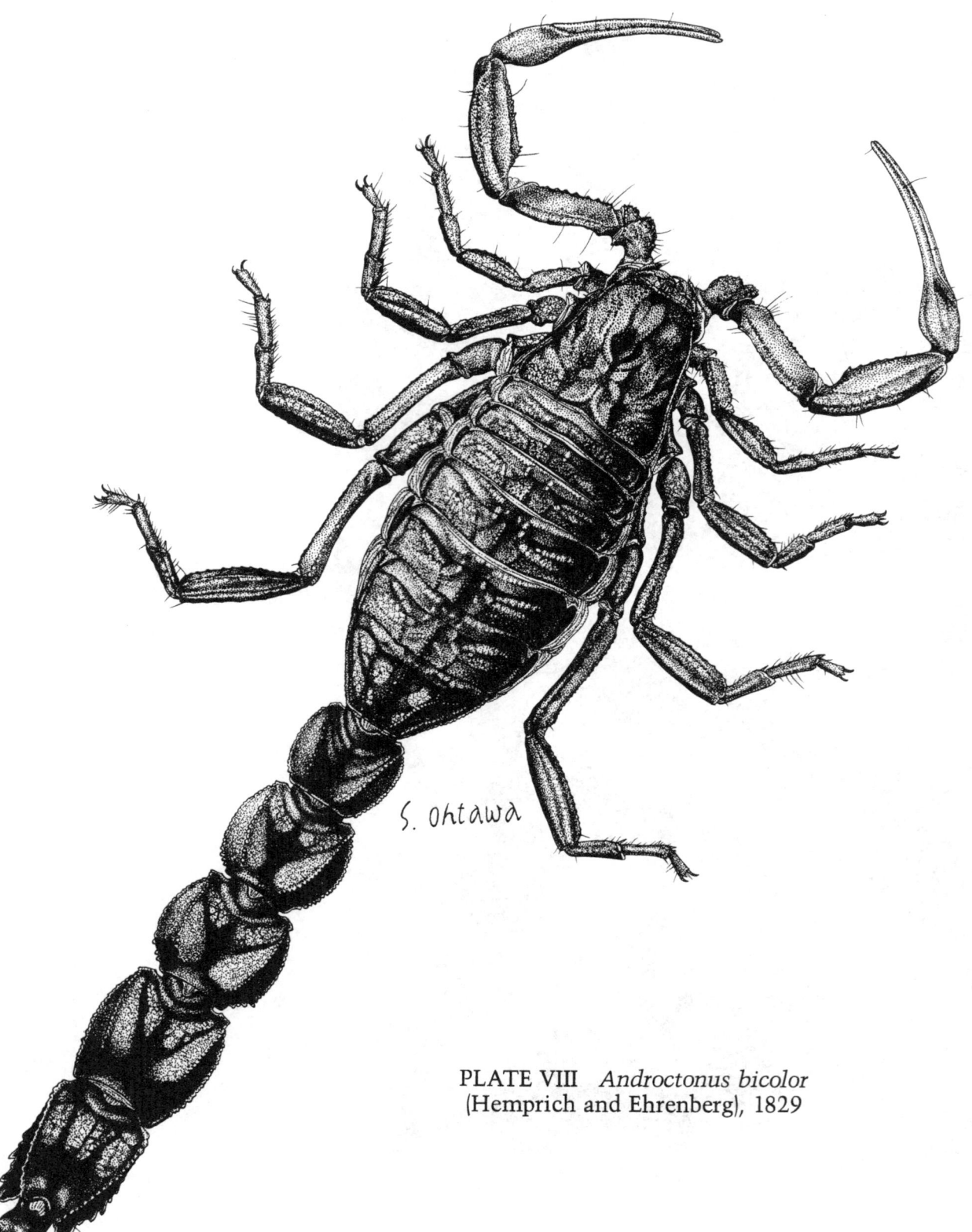

PLATE VIII   *Androctonus bicolor*
(Hemprich and Ehrenberg), 1829

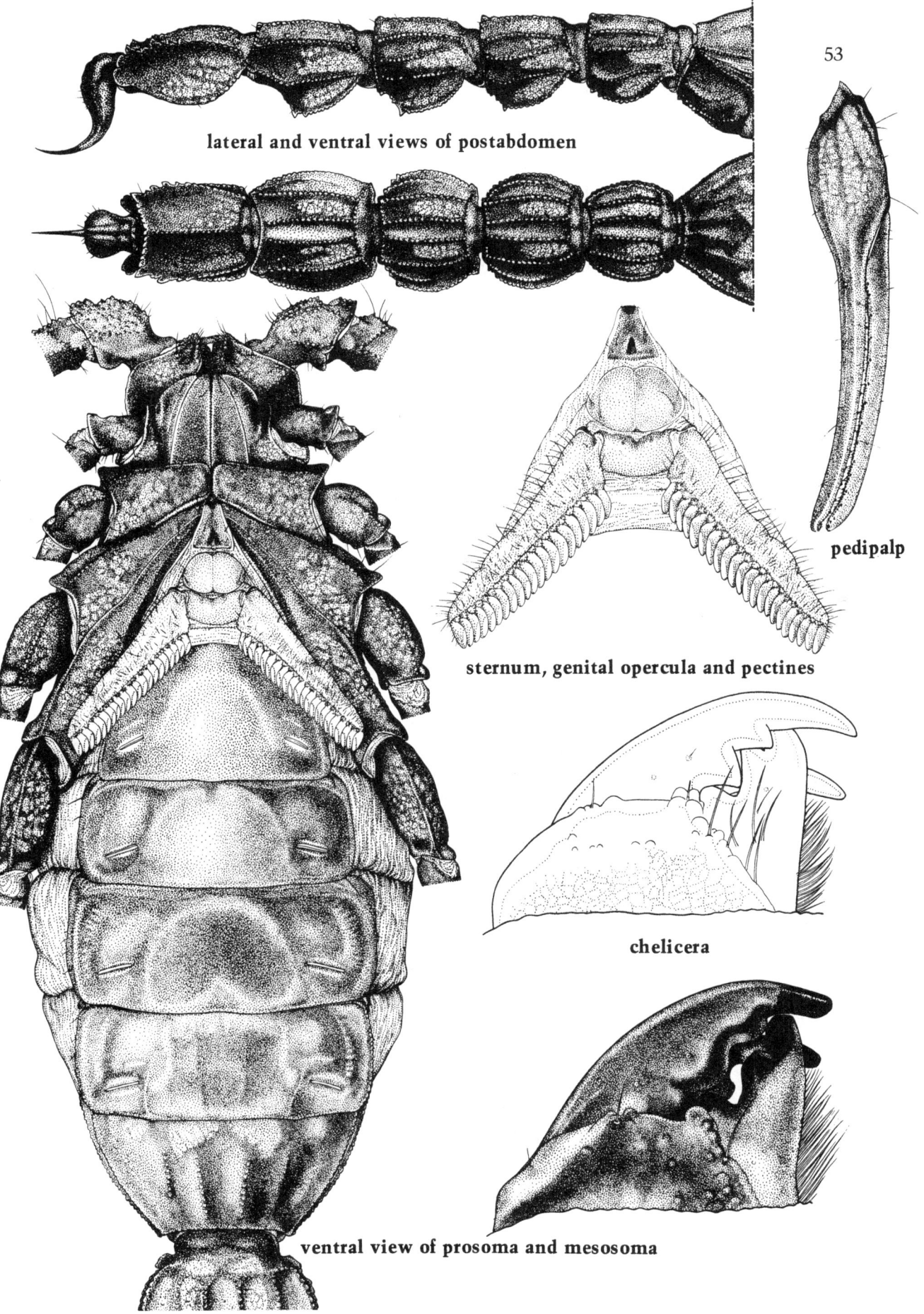

53

lateral and ventral views of postabdomen

pedipalp

sternum, genital opercula and pectines

chelicera

ventral view of prosoma and mesosoma

species in Morocco; and lastly, *A. sergenti*, Vachon, 1948 in the Atlas Mountains of Morocco.

There is virtually no information concerning potency of the venom of any of these scorpions and there are brief clinical notes on only two. Minning and Zumpt (1942), in referring to *A. aeneas liouvillei*, a subspecies of *A. aeneas* found in Morocco, termed it the "black scorpion" and stated that its venom was not as potent as that of *A. australis*, but that its sting had caused deaths among children. Levy (1965) in describing clinical aspects of envenomation by *A. mauretanicus* wrote:

> The patient is pale and often shivering. The limbs are cold and body temperature and arterial pressure may be lowered. The patient may go into shock and collapse as soon as an hour after the sting. Convulsions, delirium, diarrhea, vomiting and difficulty in respiration may precede coma and death.

No antivenins have been prepared with venoms of any of these species.

Detailed and well illustrated descriptions of each of the species of the genus except *A. bicolor* have been given by Vachon (1952).

Genus *Buthotus* Vachon, 1949

In many areas, species of this Old World genus occur in common with members of genus *Buthus*. Among characteristics common to both genera are a lyre-shaped marking formed by keels on the carapace and three longitudinal keels on the dorsal surface of the mesosoma. However, the H-shaped marking on the carapace, formed by the central medial keels, in genus *Buthus* is absent in *Buthotus*. In addition, under close examination, it can be seen that in species of *Buthotus* there are four non-linear granules just proximal to the terminal tooth of the movable finger (tarsus) of the pedipalp. Only three such granules are present in *Buthus*.

Various species of *Buthotus* occur throughout Africa, the Middle East, and Central Asia. Vachon and Stockmann (1968) have proposed that the genus has developed three "lines" of species, one in each of the geographic regions listed above. Of the approximately 20 species and subspecies that have been described only one, *Buthotus tamulus*, an Indian scorpion, has been shown to be of significant medical importance, according to Balozet (1971).

Adults of species of the genus range in length from 4–10

cm. These scorpions occur in a variety of habitats and climatic conditions. According to Vachon and Stockmann (1968) at least one African species, *B. minax*, infests dwellings in Chad.

Buthotus tamulus (Fabricius), 1828 [PLATES IV, V, IX, AND FIGURE 1]

This species, together with the five subspecies which have been described, is found throughout India. According to Pocock (1900), the color is variable, ranging from black through brown, reddish brown, and "greenish" to yellow. Among several hundred specimens obtained from a dealer in Bombay and maintained in the laboratory of the writer, the carapace and mesosoma were usually brown, with the metasoma somewhat lighter in color, and the appendages almost yellow. The ocular tubercle was dark. In some specimens, there was a longitudinal, median, dorsal, double stripe on the mesosoma. This was distinctly lighter in color than the remainder of the dorsum.

Apparently *B. tamulus* is ordinarily not a house-infesting species. Deoras (1961) in commenting on collection of several thousand scorpions from only a few villages in India, noted that *B. tamulus* was found under stones. In the laboratory, Whittemore *et al* (1963) found that scorpions of this species did well if fed on crickets and given an adequate water supply. While mating was not observed in the laboratory, many litters were produced. In a series of 15 gravid females, litter size varied from 30 to 61. In most cases, the first moult occurred three to four days following birth, and the second 36 to 40 days after birth. Following this, there was considerable variation. One specimen, observed from birth, moulted on the following post-birth days: 3, 36, 56, 91, 141, and 193. When these observations ceased, the scorpion was not yet mature. It is well known that variation in moulting in young scorpions is dependent upon availability of food and probably other environmental factors.

There is no doubt that *B. tamulus* is a dangerously venomous scorpion and that deaths, particularly among children, may follow its sting. This is well documented by the reports of Mundle (1961), Reddy *et al* (1972) and Santhanakrishnan and Balagopal Raju (1974). Clinical aspects of envenomation by *B. tamulus* are given in chapter 3. Unfortunately, no antivenin for treatment of stings by this scorpion has been produced, and treatment is entirely symptomatic.

Size range of adults of this species is usually given as 65–90 mm. The female specimen from which plate 9 was prepared was 94 mm in length. As this specimen was obtained

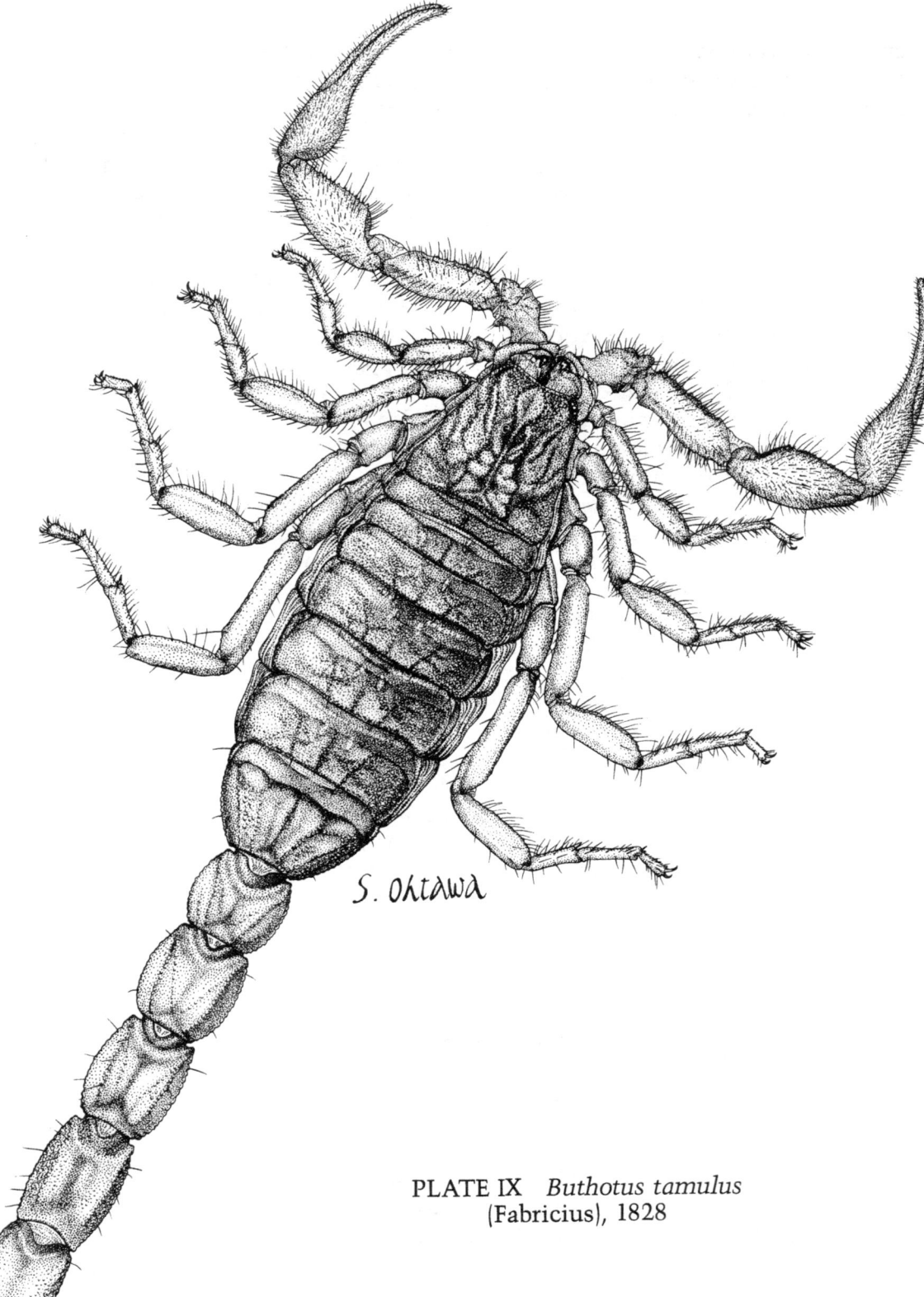

S. Ohtawa

PLATE IX   *Buthotus tamulus*
(Fabricius), 1828

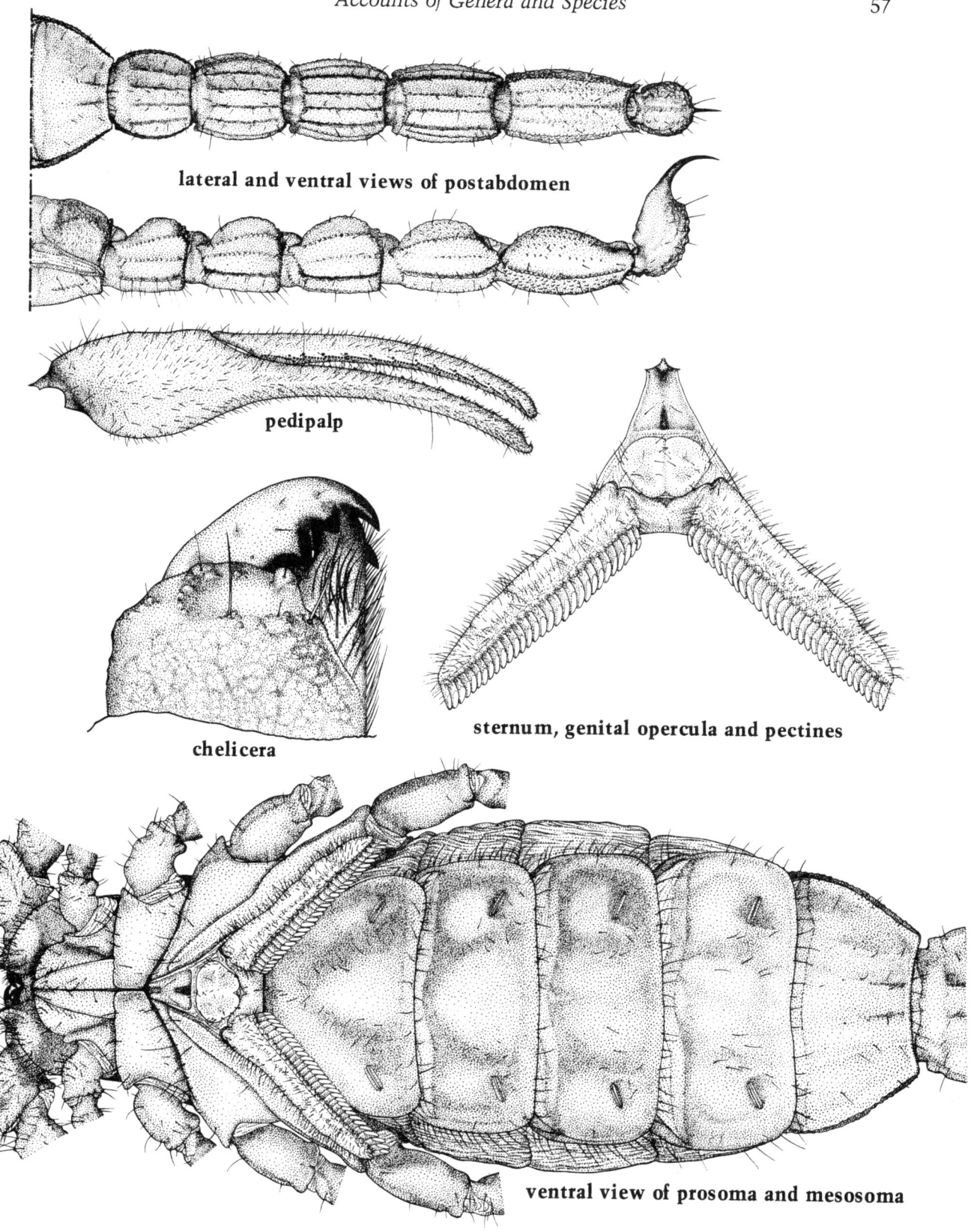

lateral and ventral views of postabdomen

pedipalp

chelicera

sternum, genital opercula and pectines

ventral view of prosoma and mesosoma

from a dealer, the locality from which it was collected is unknown. Figure 1 is a photograph of a specimen in our laboratory colony.

Three additional species of genus *Buthotus* occur in India. These are: *B. hendersoni* Pocock, 1900, *B. rugiscutis* Pocock, 1897, and *B. pachyurus* Pocock, 1897. There have been no reports of injury to man by these scorpions and nothing is known of the nature or toxicity of their venoms. No specimens were available for study.

Genus *Buthus* Leach, 1815

Although at one time over 100 species of scorpion, including all the dangerously venomous species of the Old World, were assigned to genus *Buthus*, erection of several new genera, particularly *Androctonus, Leiurus,* and *Buthotus,* considerably changed this situation. As conceived by Vachon (1952), the genus is restricted to less than 10 valid species plus several subspecies. These are distributed from southern France and Spain through the Middle East into North Africa in a variety of habitats and at altitudes ranging from sea level to mountains and high plateaus. *Buthus occitanus* (Amoreaux) 1789 which, together with its subspecies is the most widely distributed member of the genus, is also the only species of significant medical importance.

One of its most obvious generic characteristics is an "H" shaped marking on the cephalothorax formed by the central median keels. This, plus the presence of three, rather than four, non-linear granules just proximal to the terminal tooth of the movable finger of the pedipalp serves to separate species of *Buthus* from those of *Buthotus*. In this genus, and in some related genera, the lateral central keels and posterior median keels are joined to form the outline of a lyre.

Lack of both the lyre-shaped marking and the "H" on the cephalothorax serve to separate *Androctonus* from *Buthus*. In addition, the vesicle in species of *Androctonus,* is distinctly narrower than the fifth segment of the postabdomen. Five, rather than three, keels on the first and second abdominal tergites separate genus *Leiurus* from both *Buthus* and *Androctonus*.

Adults generally exceed 4 cm in length and may attain a length of more than 11 cm.

Buthus martensi Karsch, 1879 [PLATE X]

This species is mentioned and illustrated here only because Balozet (1971) stated that the species had been reported to have a "rather active toxin." Balozet gave the distribution of the species as Korea and Manchuria, while Kraepelin (1899) wrote that it was found in Mongolia and China.

In living specimens, the preabdomen is brown to blackish and the postabdomen yellow, with the exception of the fifth segment, which is darker. The pedipalps and the vesicle of the telson are yellow. The species is said to attain a length of 60 mm. The specimen illustrated in plate 10 was 54 mm in length. It, and several others in the same lot, were collected at Peking, China in 1922 by F. R. Wulsin. These specimens (acc. no. 85337 in the collections of the United States Museum) were identified by E. A. Chapin in 1949.

Buthus occitanus (Amoreux), 1789 [PLATE XI]

This widely distributed scorpion, represented by several subspecies and a number of varieties, is found in southern France, Spain, Italy, Greece, the Balkans, and several of the Mediterranean islands. One subspecies occurs in Israel and Jordan, while others are found in Egypt and throughout much of North Africa, particularly in Algeria, Morocco, Tunisia, and Chad. A detailed account of the North African subspecies has been given by Vachon (1952).

Among the various subspecies, color ranges from clear straw yellow to rusty brown, occasionally with distinct dark bands on the preabdomen. The fifth segment of the postabdomen may be darker than the other. The keels of the cephalothorax are well formed and distinct, especially in the male. However, the lateral posterior keels are often obscured by granulations. The aculeus of the telson is of variable length, but usually does not exceed the length of the vesicle. Tergites of the preabdomen possess three longitudinal keels, although these are not always distinct. The seventh sternite has four granular keels in both sexes. There are 10 keels on the first segment of the postabdomen and 8 on the second and third segments. The lateral ventral keels of the fifth segment possess distinct, uneven teeth which may vary in number and relative size among the subspecies. Pectinal teeth of the female vary in number from 19 to 30; those of the male from 25 to 36. Adults usually vary in length from 4.5 cm to 7.5 cm.

Balozet (1971) commented that the toxicity of *B. occitanus* in Northwest Africa appeared greater than in Europe. Bouisset and Larrouy (1961) also wrote that, while no case of any severity from the sting of *B. occitanus* had been reported from metropolitan France, it was otherwise in Algeria where fatal cases had been seen at the Pasteur Institute. They found that small children accounted for about a quarter of the cases of

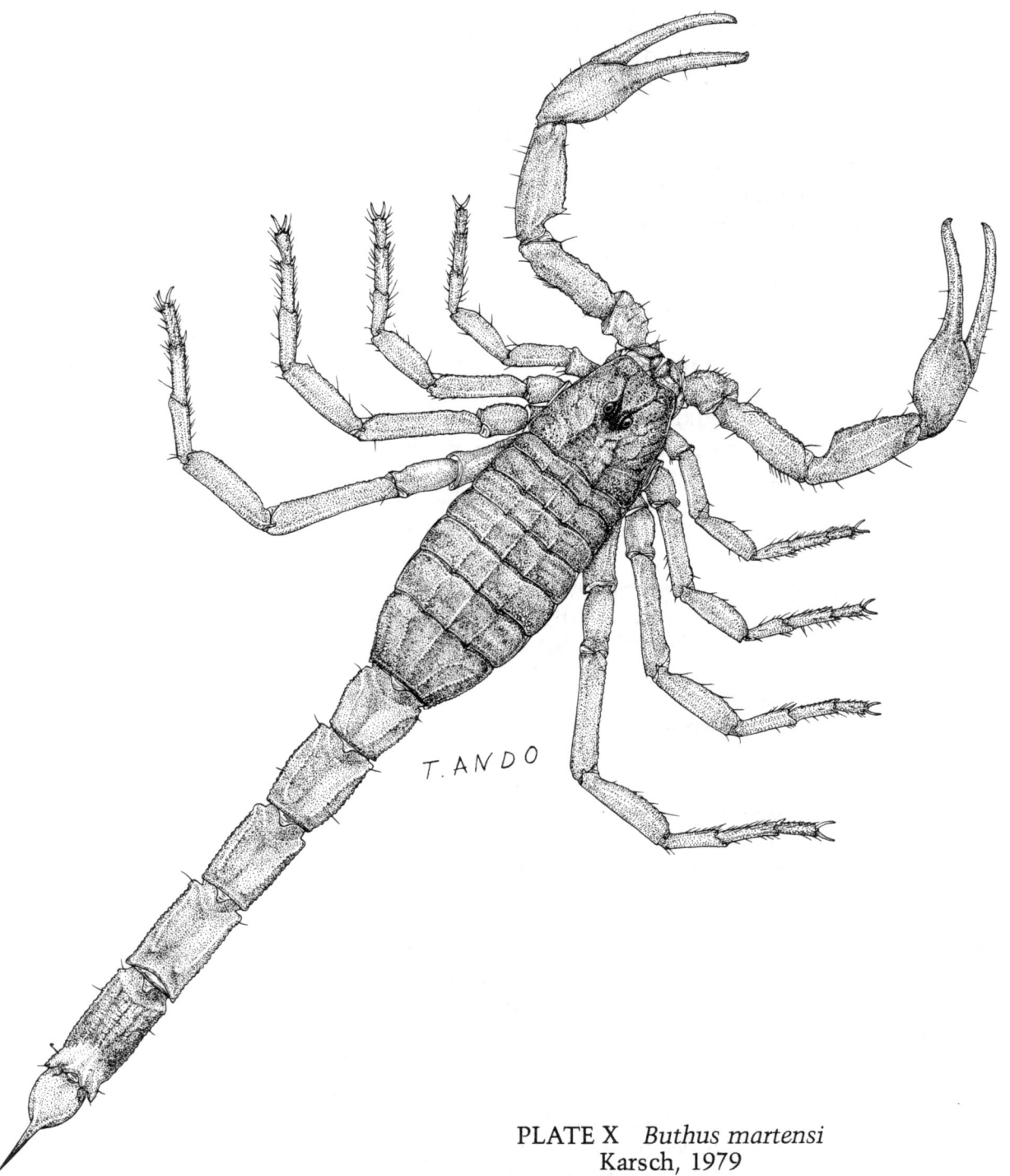

PLATE X *Buthus martensi*
Karsch, 1979

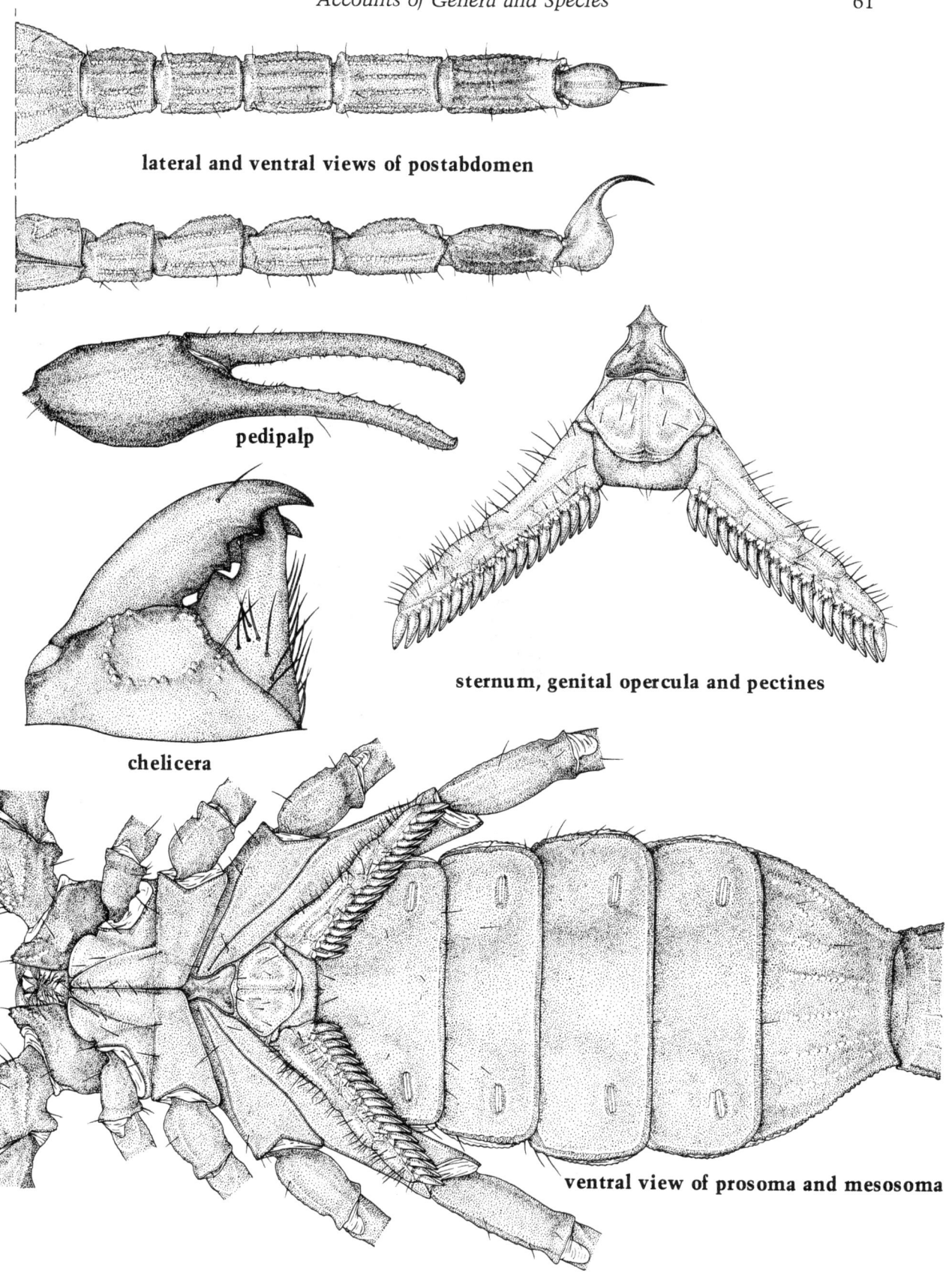

lateral and ventral views of postabdomen

pedipalp

chelicera

sternum, genital opercula and pectines

ventral view of prosoma and mesosoma

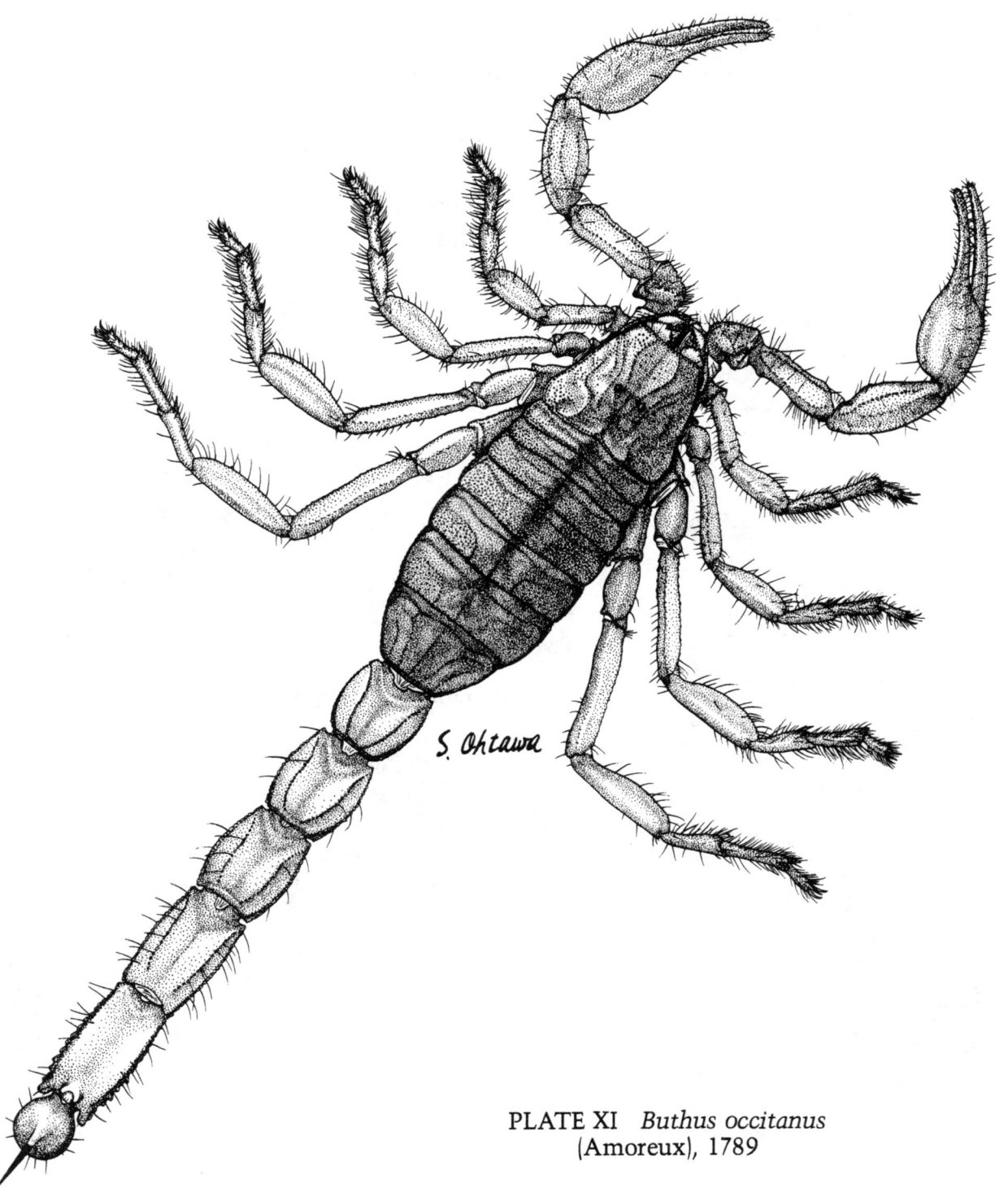

PLATE XI *Buthus occitanus*
(Amoreux), 1789

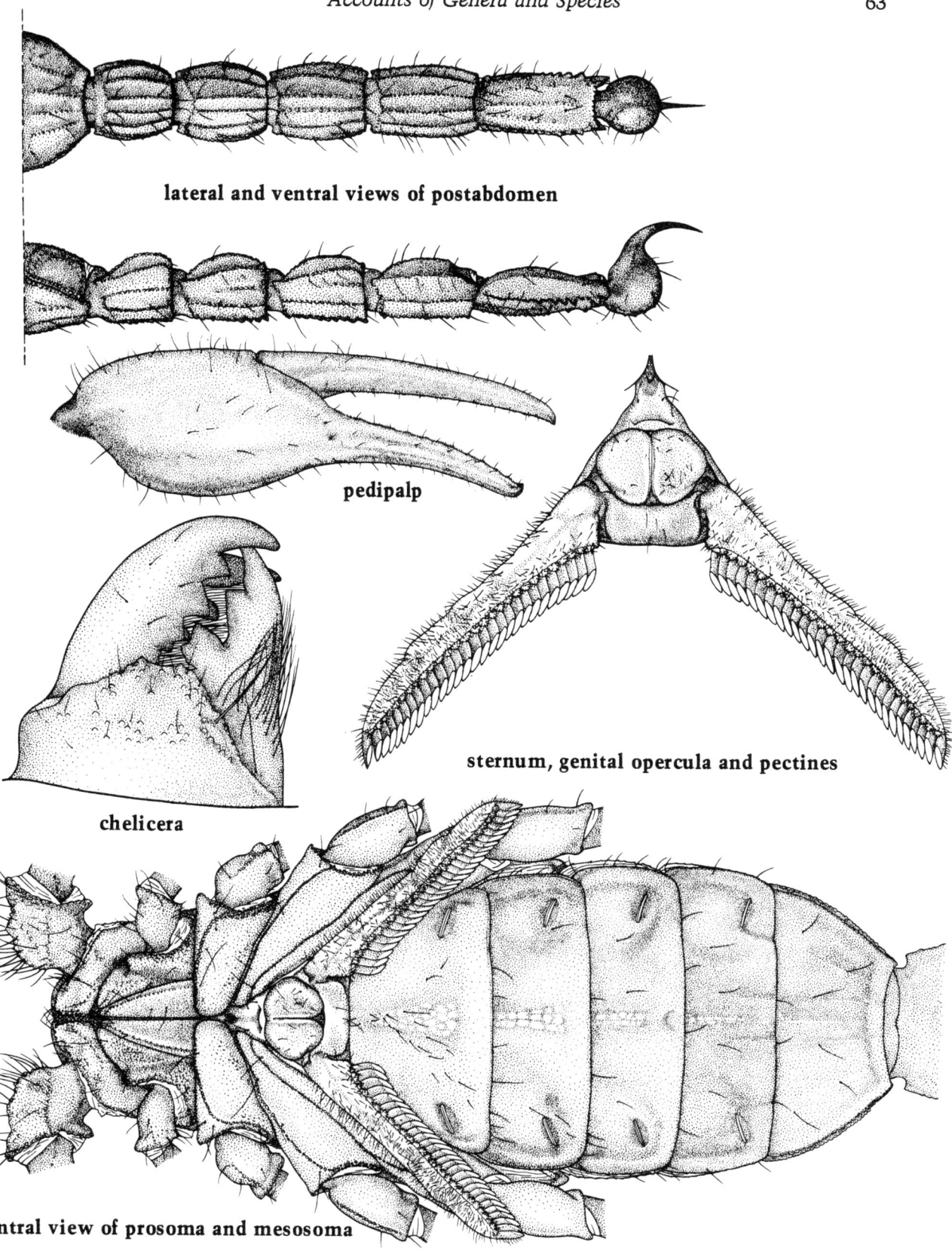

lateral and ventral views of postabdomen

pedipalp

chelicera

sternum, genital opercula and pectines

ventral view of prosoma and mesosoma

scorpion sting seen by them, and that most of these occurred during the hot season from June to September. Usual symptoms were intense pain, localized at the area of the sting, and slight local edema. General symptoms were those of shock. Wahbeh (1965) reported five deaths among 28 children less than 10 years of age who had been stung by *B. occitanus* in Jordan.

Bouisset and Larrouy found that an antivenin prepared at the Pasteur Institute in Algiers for treatment of stings by *Androctonus australis* was equally effective in treatment of envenomation by *B. occitanus*. Additional examples of antivenin paraspecificity were reported by Whittemore *et al* (1961). It was found that an antivenin produced with venom of the Turkish scorpion, *Androctonus crassicauda* was highly effective, in neutralization tests using mice, against venom of *B. occitanus*. It was also discovered that an antivenin produced in Mexico against venom of scorpions of genus *Centruroides* was as effective as a homologous antivenin against *B. occitanus* venom.

Rolli (1972) found that 25% BHC and 1% lindane in powder formulations gave 100% kills in laboratory tests with *B. occitanus*. Field tests with 1% lindane were also described as effective.

Mating, birth and early activities of newly born scorpions were described by Fabre (1911) in a rather anthropomorphic, but humorous and entertaining manner.

The specimen figured on plate 11 was 64 mm in length. This, and other examples of the species, were gifts of Dr. Lucien Balozet who, at the time (1960), was on the staff of the Institut Pasteur d'Algerie, Alger. Collecting data were not given.

Genus *Leiurus* (Hemprich and Ehrenberg), 1829

The genus *Leiurus* was originally designated as a subgenus of *Androctonus*. Although several species were at first assigned to it, it was later found that all but one of these, the genotype, *L. quinquestriatus*, more properly "belonged" in other genera. In raising *Leiurus* to full generic status, Vachon (1952) stated that the distinguishing generic characteristic was the presence of five, rather than three, keels on the first two abdominal tergites. The single species assigned to the genus is found in the Middle East and North Africa.

Leiurus quinquestriatus (Hemprich and Ehrenberg), 1829 [PLATE XII]

This dangerously venomous scorpion is found from Turkey southward through Lebanon, Syria, Jordan and Israel into

North Africa, where it constitutes a medical problem particularly in Egypt, Libya, Sudan, and Algeria. Subspecies have been described, although Vachon (1952) wrote that without further study validity of these must remain in question. *L. quinquestriatus* is found at various altitudes in arid, often semi-desert areas. Shulov (1962) noted that in Israel it is found mainly in hill regions under stones and that in such situations the color of these scorpions varies from light yellow to orange red, according to the environment. This, of course, is not unusual, and is seen in many desert-dwelling vertebrates and invertebrates. Minning and Zumpt (1942) described North African specimens as yellow in color while Vachon (1952) mentioned wine colored specimens from one area of Algeria. In some specimens, the fifth post-abdominal segment is darker than the appendages or remainder of the body.

Although *L. quinquestriatus* is not a house-infesting species, it constitutes a definite public health problem in each country where it is found. This is in large part because stings often occur in remote areas where prompt medical attention is not available. Tulga (1960) found that in laboratory tests using rats, *L. quinquestriatus* venom was from four to five times more toxic than that of *Androctonus crassicauda*. Whittemore *et al* (1963) found that the average venom yield (when dried) in a series of 192 venom collections from specimens of *L. quinquestriatus* from Israel was 0.483 mg. Whittemore and Keegan (1963) in an attempt to compare toxicity of venoms devised the term *toxicity factor* i.e. the number of LD 50's/ average venom yield (upon electrical stimulation of the scorpion) as determined by i.p. injection into white mice. On this scale, the obtained value of 18.6 was less than that found in studies with three much smaller species of the genus *Centruroides* from Mexico. Both Balozet (1971) and Shulov (1962) emphasized the severe clinical aspects of envenomation. These are indicated in table 1 in chapter 2 of this publication. In laboratory studies, Tulga (1960 and 1964) found that an antivenin produced with venom of *Androctonus crassicauda* was effective in neutralizing venom of *L. quinquestriatus* even though it was ineffective against venom of the homologous species. Formerly, at least, a *L. quinquestriatus* antivenin was produced at the State Serum and Vaccine Institute, Cairo, Egypt.

Adults of *L. quinquestriatus* may reach a length of 90 mm or more. The specimen illustrated on plate 12 was among a group of living specimens sent from the Hebrew University of

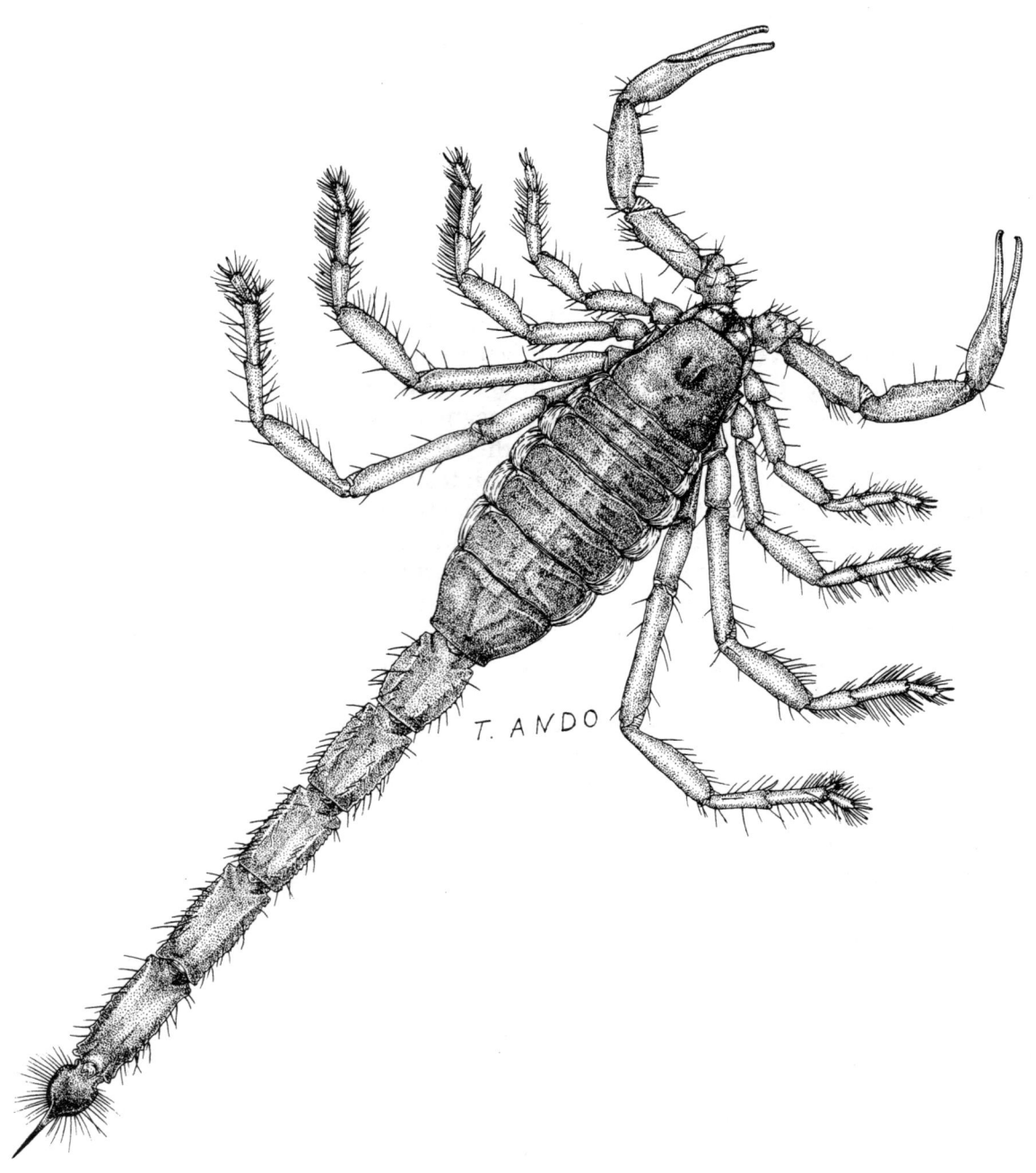

PLATE XII *Leiurus quinquestriatus*
(Hemprich and Ehrenberg), 1829

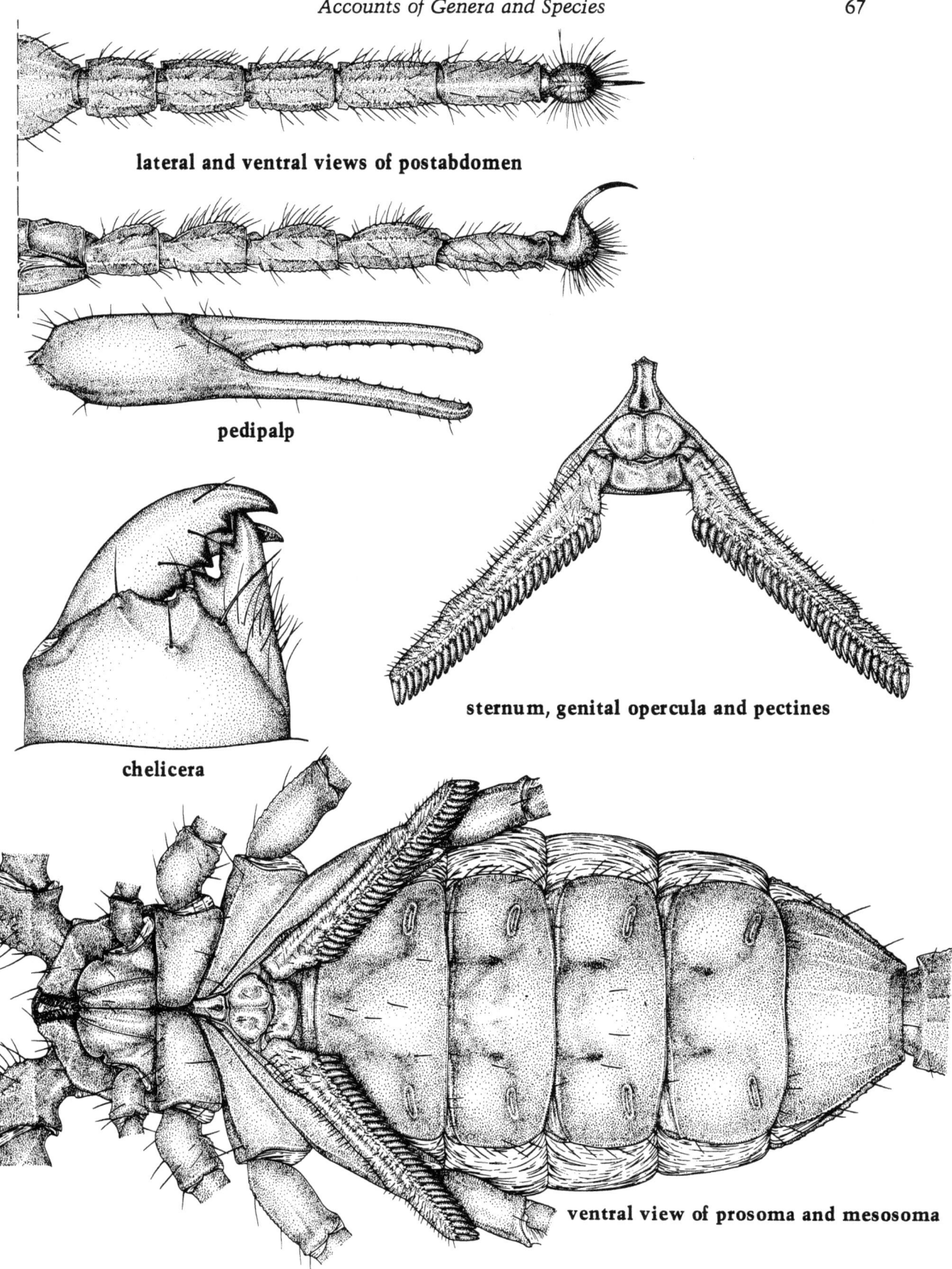

lateral and ventral views of postabdomen

pedipalp

chelicera

sternum, genital opercula and pectines

ventral view of prosoma and mesosoma

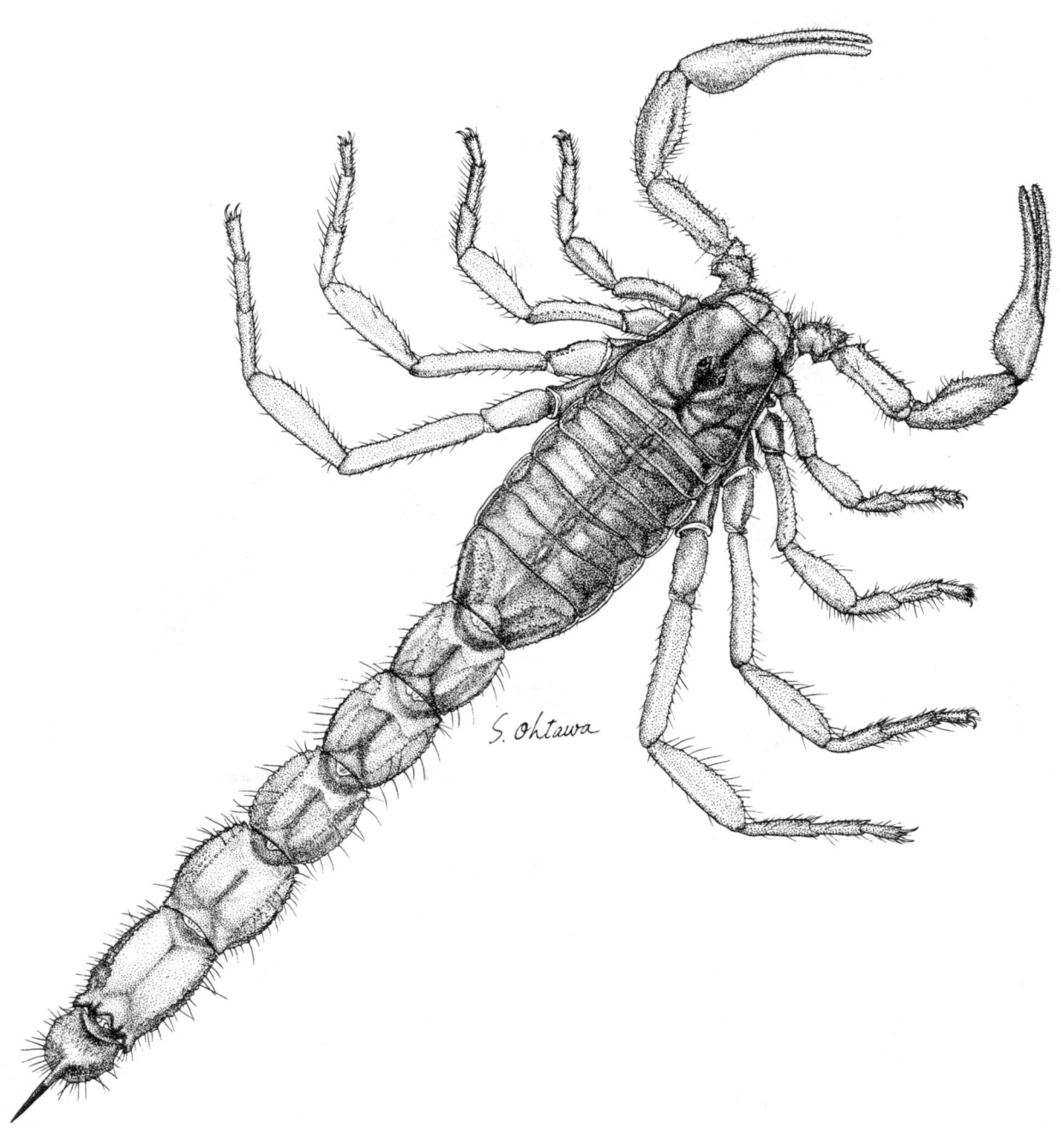

PLATE XIII   *Parabuthus triradulatus*

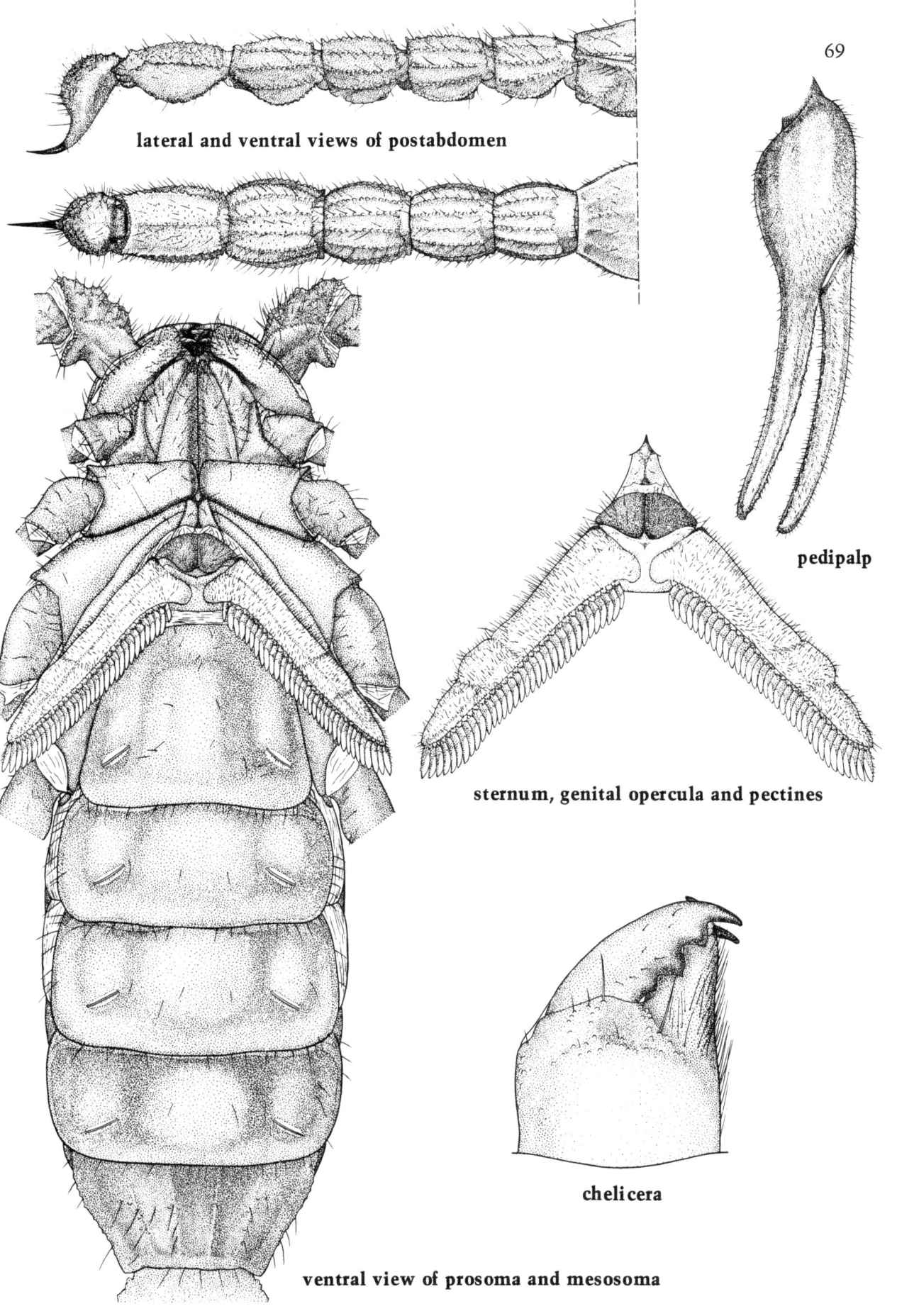

lateral and ventral views of postabdomen

69

pedipalp

sternum, genital opercula and pectines

chelicera

ventral view of prosoma and mesosoma

Jerusalem by Professor Aharon Shulov and maintained in the laboratory at Brooke Army Medical Center, Fort Sam Houston, Texas. Collecting data are not known. The specimen illustrated was 83 mm in length.

Genus *Parabuthus* Pocock, 1890

Of the approximately 30 species of genus *Parabuthus* all but a few are found only in South Africa. Vachon (1966) and Probst (1976) have reported *P. liosoma* (Hemprich and Ehrenberg) 1829 from Arabia, and Probst also reported this species from Tanzania. He also gave collection localities for *P. pallidus* Pocock, 1890 from Kenya and Tanzania. According to Stahnke (1972), species of the genus are similar to those of *Buthotus* and *Buthus* in lack of a subacular tooth, but differ from both in that there is only a single, median keel on the tergites of the mesosoma. Adults may attain a length of over 7 cm.

While stings by members of this genus usually result only in severe local pain, systemic effects and deaths have been reported, particulary in children. Generalized "stiffness" of the limbs is apparently common and, in one case, that of a seven-year-old child, generalized paralysis and death due to respiratory paralysis occurred six hours following the sting. An antivenin for treatment of envenomation by species of genus *Parabuthus* is prepared at the South African Institute for Medical Research, Johannesburg. A key to the South African species has been published by Hewitt (1918). Probst (1976) has given descriptions of *P. liosoma* and *P. pallidus*.

The specimen of *Parabuthus triradulatus* illustrated on plate 13 was loaned by Dr. A. J. Hesse of the South African Museum, Capetown.

Genus *Tityus* C. L. Koch, 1836

Species of this South American genus are similar in size, habits, and general appearance to scorpions of the genus *Centruroides*. Unlike the condition in *Centruroides*, there are no supernumerary granules flanking the oblique, median rows of granules on the movable finger of the pedipalp. In his key to the Buthidae, Stahnke (1972) noted that there were not less than 14 oblique rows of median granules in species of *Tityus* and not more than 9 in scorpions of genus *Centruroides*.

Of the approximately 30 species assigned to the genus several are dangerously venomous. These are: *Tityus serrulatus* Lutz and Mello-Campos, 1922 in Brazil; *T. bahiensis* (Perty) 1922 in Brazil and Argentina; *T. trinitatis* Pocock, 1897 in Trinidad and Venezuela; *T. trivittatus* Kraepelin, 1898 in

Argentina, Paraguay, Uruguay and Brazil; *T. trivittatus char-reyroni* Vellard, 1932 in Brazil; and *T. cambridgei* Pocock, 1897 in Guyana and perhaps, as exemplified by *T. trinitatis,* are listed in Table 1 in chapter 2 of this publication.

Although Ewing (1928) listed two species of the genus, *T. floridanus* Banks, 1900 and *T. tenuimanus* Banks, 1900, as occuring in the United States, Muma (1967) found no examples of *T. floridanus* in several years of extensive collecting in Florida and believes that the single record from that state must have been in error. *T. tenuimanus* too has been reported only once from Buena Vista, California. The writer has been unable to find any additional information concerning the species.

Examples of only two species, *T. serrulatus and T. bahien-sis* were available for study. These, illustrated in plates 14 and 15, were sent through the courtesy of Dr. Wolfgang Bücherl of the Instituto Butantan, São Paulo, Brazil.

Tityus serrulatus
Lutz and Mello-Campos,
1922 [PLATE XIV]

This species, generally considered to be the most dangerous scorpion in Brazil, occurs only in that country, where, according to Bücherl (1971), it is found in the states of Sergipe, Bahia, Minas Gerais, Espirito Santo, Rio de Janeiro, Goiás and São Paulo.

In life, the cephalothorax and all but the last segment of the preabdomen vary in color from yellow to blackish brown with some dark marking on each segment. The last preabdominal segment, the postabdomen and the appendages are pale yellow, except that the fingers of the pedipalp and the telson are reddish brown. The aculeus may be almost black. An important taxonomic characteristic is that the granules of the dorsal median keel on the third and fourth segments of the postabdomen are distinctly enlarged to form a serrated appearance when viewed laterally. This is shown on plate 14.

The medical importance of *T. serrulatus* stems from the toxicity of its venom plus its habits, which are similar to those of *Centruroides sculpturatus* in the United States. As a "domestic" species it is a frequent cause of scorpion sting. In some parts of Brazil it is exceedingly common. Bücherl (1971) wrote that during the period 1949 to 1963 about 138,000 specimens of *T. serrulatus* were captured in four cities of Brazil and sent to the Instituto Butantan for venom research and antivenin production. In the areas where the species is found, various insecticides have been tested in control programs. Souza *et al* (1954) found lindane to be effective in scorpion

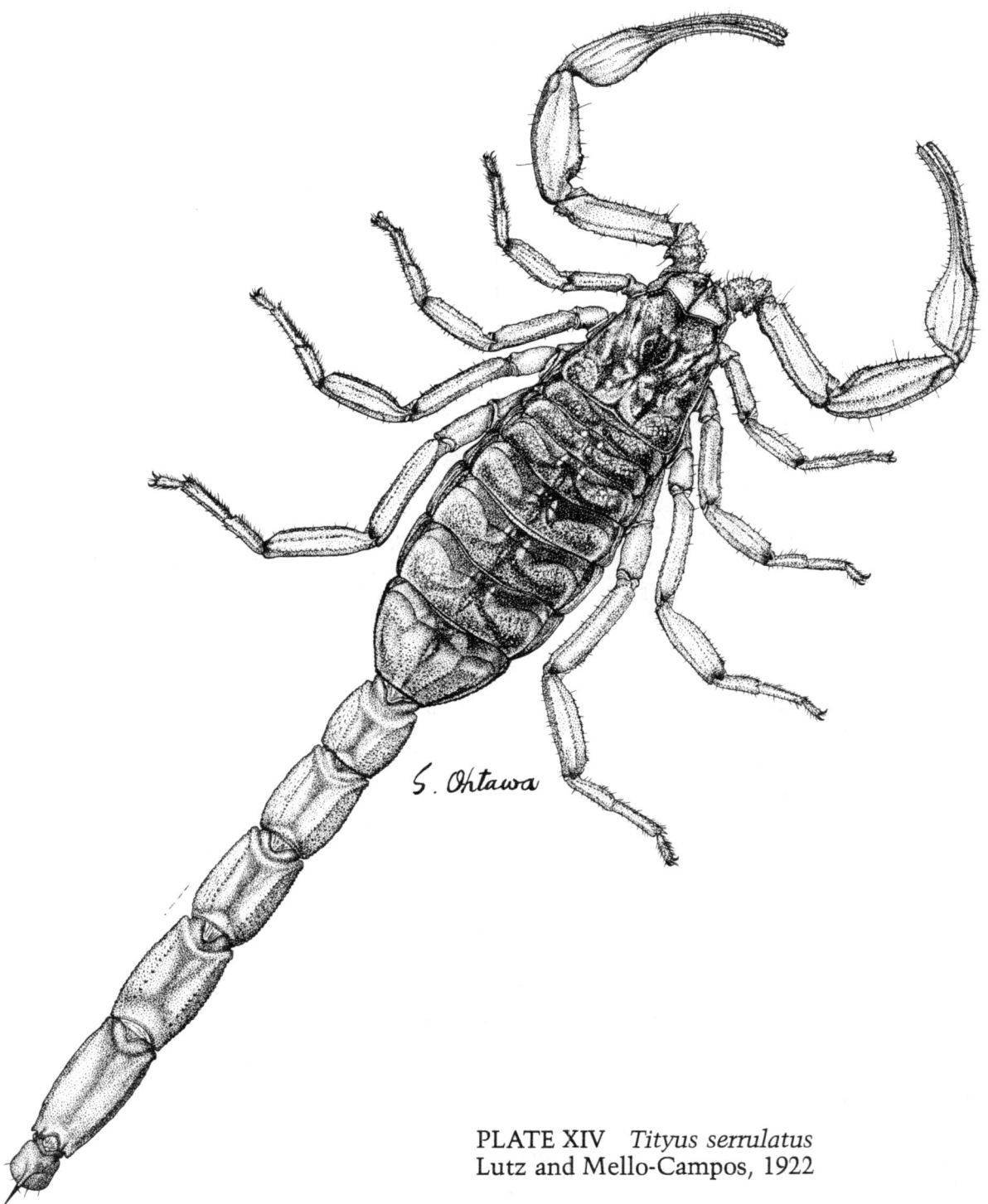

S. Ohtawa

PLATE XIV   *Tityus serrulatus*
Lutz and Mello-Campos, 1922

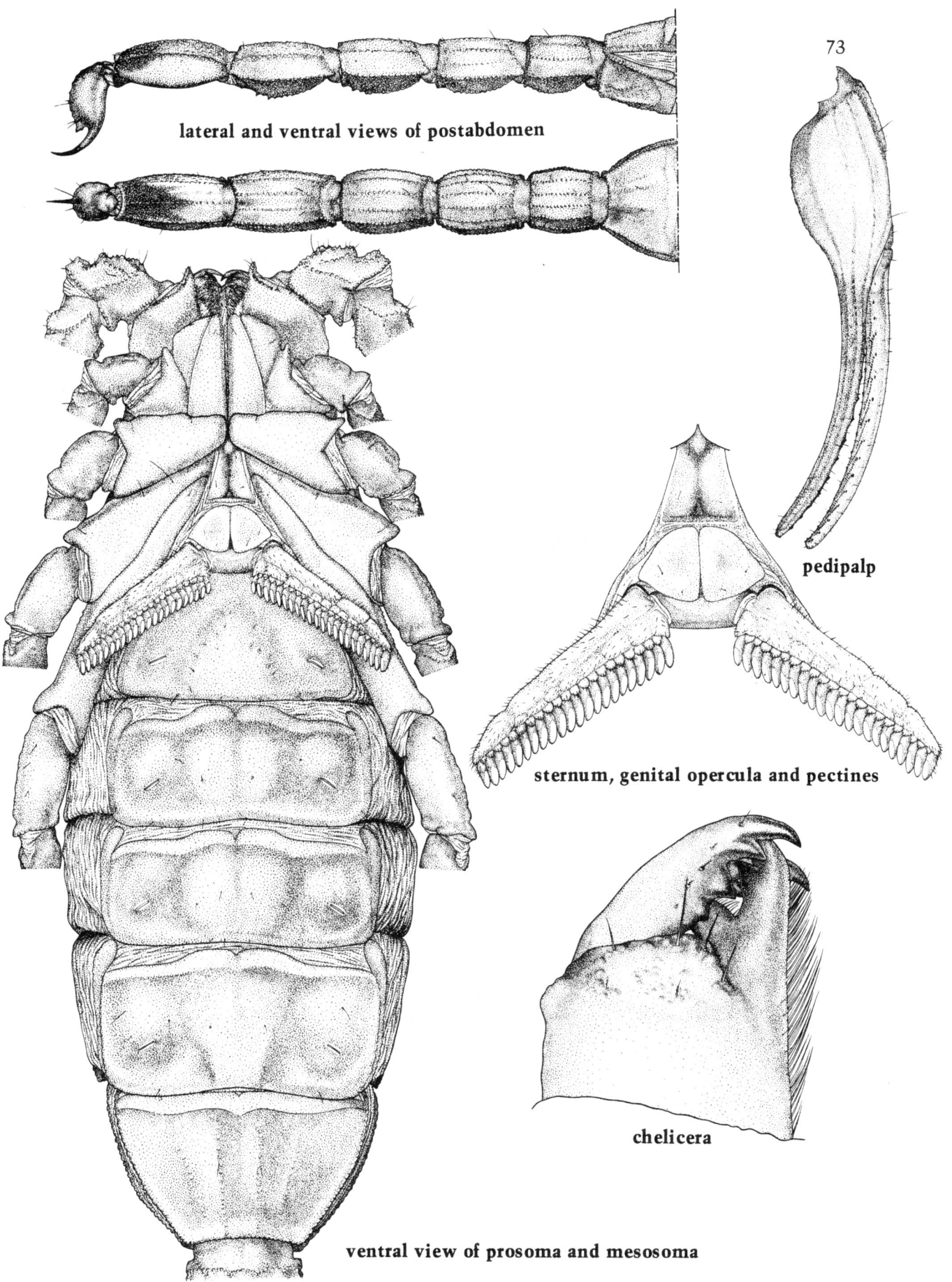

lateral and ventral views of postabdomen

73

pedipalp

sternum, genital opercula and pectines

chelicera

ventral view of prosoma and mesosoma

control when it was used as a residual spray in and around buildings. Bücherl (1971) also found lindane to be effective but noted that contact with this insecticide excited the scorpions, making them more active and increasing the possibility of envenomation. Because of this, in later years, a 5% DDT emulsion was used in control programs.

Tityus serrulatus is unique in that, apparently, reproduction is entirely by parthenogenesis. A male of the species has never been found. Parthenogenetic birth in the laboratory was first described by Matthiesen (1962) and again by San Martin and Gambardella (1966). Bücherl (1971) found only females among over 60,000 specimens received at his laboratory. In addition, he found only ovaries in about 8,000 dissected specimens.

Adults of this species may attain a length of about 70 mm. The scorpion figured on plate 14 measured 58 mm. This specimen, and others so kindly supplied by Dr. Wolfgang Bücherl, were from the colony maintained at the venomous animals laboratory at the Instituto Butantan, São Paulo, Brazil.

Tityus bahiensis
(Perty) 1922 [PLATE XV]

According to Bücherl (1971), this species, considered second only to *T. serrulatus* in medical importance in Brazil, is found in Argentina in the province of Buenos Aires and in Brazil in the following states: Bahia, Espirito Santo, Minas Gerais, Goiás, Rio de Janeiro, São Paulo, Santa Catarina, and Mato Grosso.

In this scorpion, the cephalothorax and preabdomen are uniformly brown and the postabdomen and appendages reddish brown, except that the patella of the pedipalp is dark brown or black. In addition, the species differs from *T. serrulatus* in that the granules of the dorsal median keel on the postabdominal segments three and four are not enlarged and do not present a serrated appearance from a lateral view.

Like *T. serrulatus* this is a house-infesting scorpion. However, it differs from *serrulatus* in that both sexes exist. Length of adults may vary from 60–75 mm. As in many other species, the postabdomen of the male is longer than that of the female. The female specimen figured on plate 15 was 68 mm in length. Specimens for study were obtained from Dr. Wolfgang Bücherl of the Instituto Butantan, São Paulo, Brazil.

In addition to *T. serrulatus* and *T. bahiensis*, two less venomous house-infesting species occur in Brazil. These are the widely distributed *T. trivittatis trivittatis* and *T. trivittatus charrenyroni* a species which, according to Lourenco (1975), is the most frequently encountered scorpion in the Distrito Federal. Both are yellow or yellowish red dorsally with three, sometimes obscure, dark stripes. While this writer has not examined a scorpion of either species, according to Mello-Leitao (1945) the fifth postabdominal segment is darker than the others in *T. t. trivittatus*, while in *T. t. charrenyroni* the fourth and fifth segments are of the same color. In both, there are black marks on the segments of the pedipalps and legs. Bücherl (1971) states that these are more prominent in young specimens. Although no specific information is available concerning potency of the venom of these scorpions, Abalos (1963) wrote that stings by *T. t. trivittatus* in Argentina caused intense, local, burning pain, anxiety, and sometimes difficulty in swallowing, but these symptoms were transitory and subsided within a few hours.

Tityus cambridgei is a blackish, forest-dwelling scorpion which, according to Mello-Leitao (1945), is especially common in the Amazon basin in Brazil. It has also been reported from Surinam, Ecuador, Guyana and Panama. No specific information is available concerning venom of this species, and the single report concerning possible medical importance is that of Floch *et al* (1950). Symptoms observed in three patients who had been stung by specimens of *T. cambridgei* included: local pain followed by profuse sweating and cyanosis, hypersensitivity, salivation, and vomiting. One patient complained of numbness in the legs, contraction of the jaw muscles and limbs, and later, convulsions, syncope, and semi-coma. He was given an injection of physiological saline, snake antivenin, and camphorated oil. Although the treatment mentioned probably had little curative effect, the patient recovered rapidly, but suffered from vertigo and was not really well for five days. This report indicates that the habits of this scorpion, which do not bring it into frequent contact with man, may be responsible for the lack of information concerning it.

Tityus trinitatis, an extremely dangerous scorpion, apparently occurs only in Trinidad and in Venezuela. The carapace and pre and postabdomen are yellowish brown to dark brown, while the legs and pedipalps are reddish and yellow. In juveniles, blackish spots may be present on the legs. Unlike *T. serrulatus* and *T. bahiensis* this is usually not a house-infesting species. In Trinidad, from where most of the information concerning effects of stings by *T. trinitatis* has come, cane

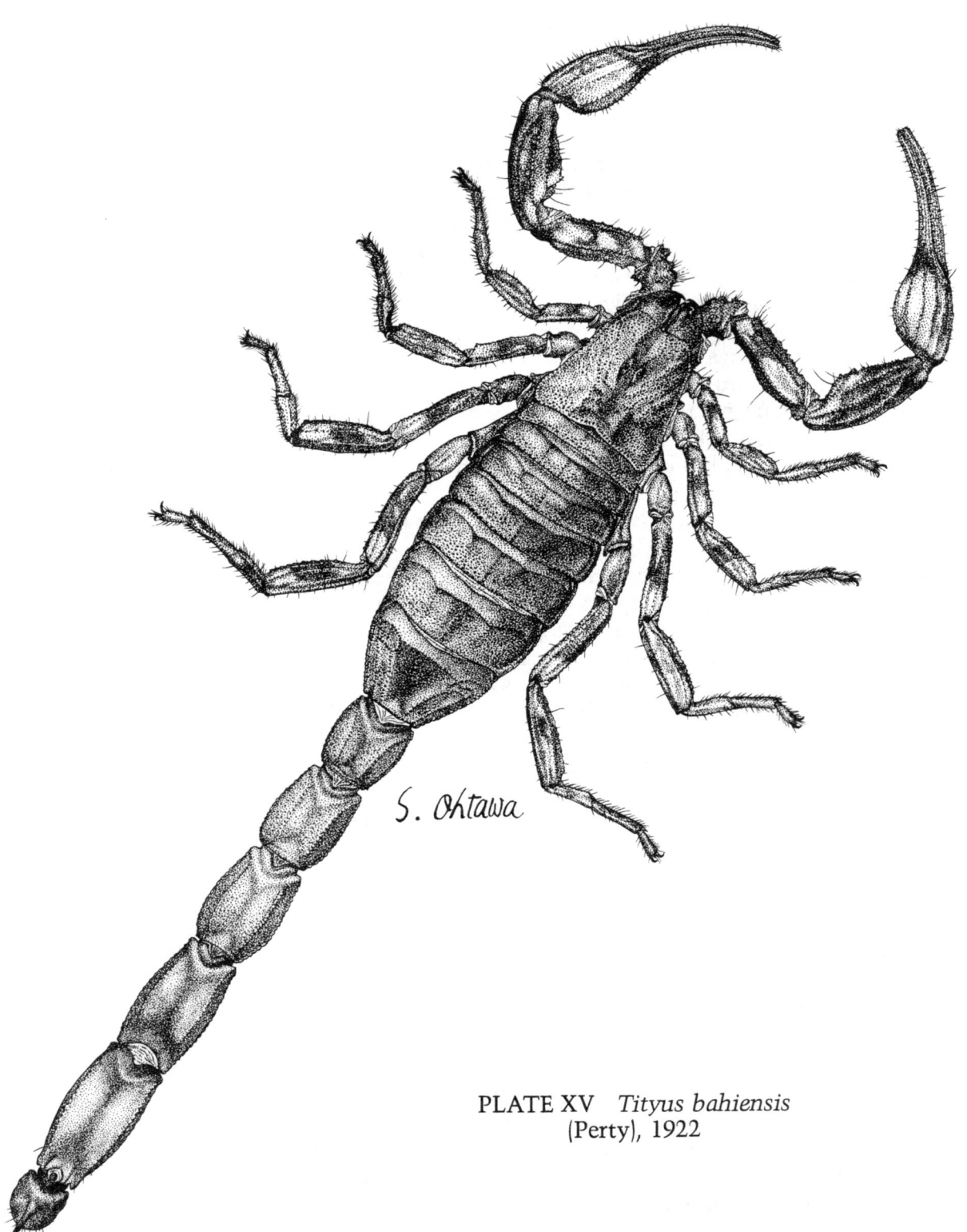

PLATE XV   *Tityus bahiensis*
(Perty), 1922

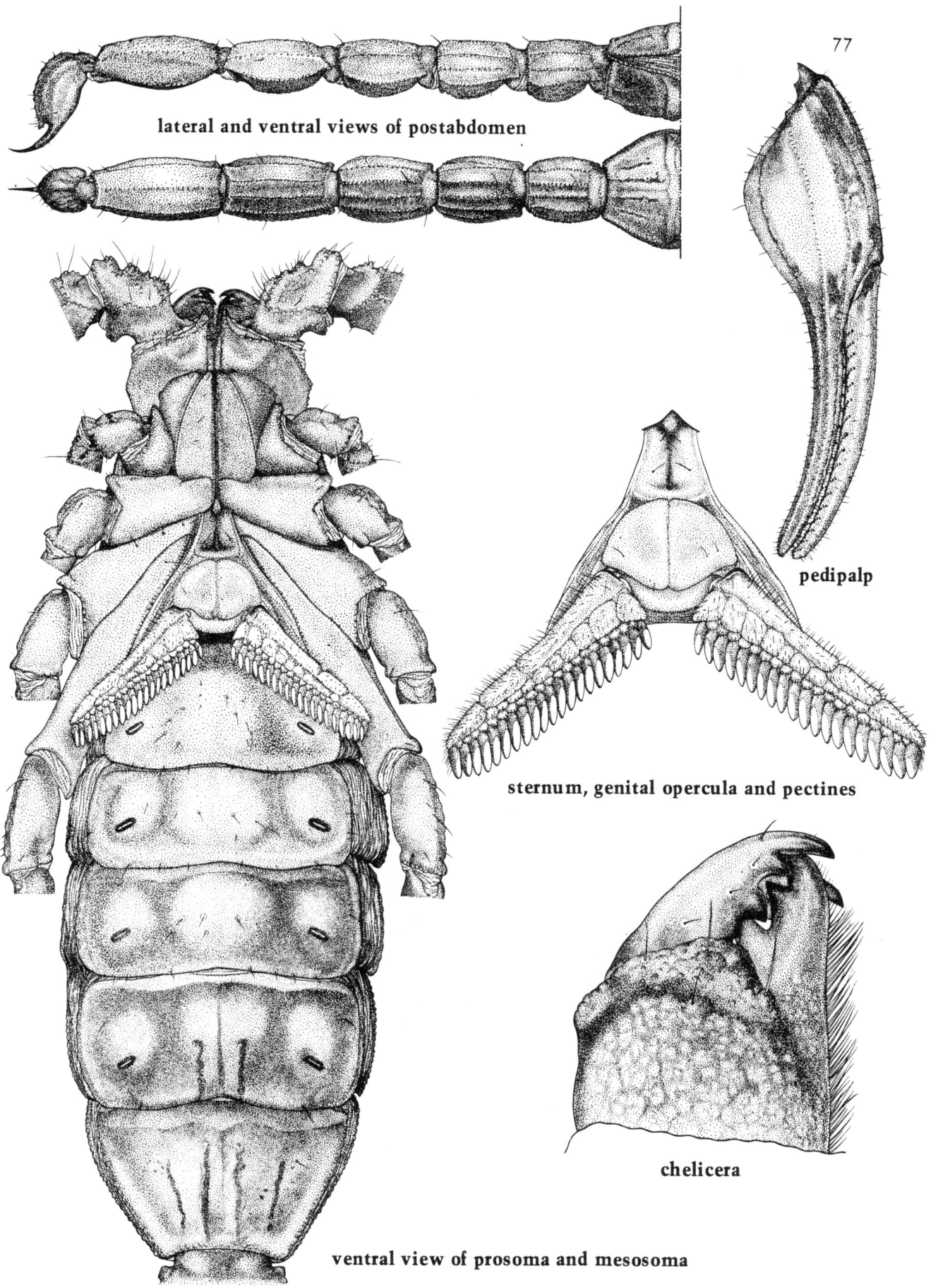

lateral and ventral views of postabdomen

77

pedipalp

sternum, genital opercula and pectines

chelicera

ventral view of prosoma and mesosoma

fields and coconut groves are cited as primary habitats for the species.

Both laboratory and clinical reports indicate that *T. trinitatis* venom is about equal to that of *T. serrulatus* in toxicity. Detailed reports by Poon-King (1963), and by Waterman (1938a, 1938b, 1950a and 1950b) describe the varied effects of envenomation and success of therapy. The clinical aspect of envenomation by *T. trinitatis* is reviewed in detail in chapter 2 of this publication.

Genus *Centruroides* Marx, 1899

Species of this New World genus occur in the United States, Mexico, Central America, the West Indies, and in several countries in South America. The "center of distribution" for the genus appears to be Mexico. While several species are dangerously venomous, stings by others result in only transient pain. In this genus, the metasoma of the male is usually much longer and more slender than that of the female. The lower margin of the fixed finger of the chelicera bears one large tooth, while that of the movable finger possesses two teeth. Tibial spurs are lacking, but there are well developed exterior and interior pedal spurs. The surface of the sternite of the first abdominal segment (just posterior to the genital operculum) is smooth or, at most, weakly granular. A dorsal furrow is often lacking on the fifth segment of the metasoma. Such a furrow, if present, is shallow. The nature and arrangement of the granular "teeth" on the movable finger of the pedipalp is of particular importance in separation of species of the genus from one another and from species of closely related genera. In members of genus *Centruroides* there are varying numbers of median oblique rows of granules which are flanked by larger, dentate granules between which are found small supernumerary granules. These can be seen on plates 16–22. (In examining these drawings a hand lens may be helpful, however, the figures *are* accurate.)

Like many other buthid scorpions, species of genus *Centruroides* have often been referred to as "bark scorpions" as they are so frequently found under loose bark on trees and in crevices in dead trees and logs. This same tendency to hide above the ground, rather than burrow, has made them "domestic" species in many areas, where they may thrive in large numbers in lumber piles, bricks, and other debris, and also enter homes where they secrete themselves in any situation offering darkness and close contact. Stahnke (1956) listed a variety of habitats, both indoors and outdoors, in which he had found specimens of the dangerously venomous *C. sculpturatus* in Arizona. Characteristically, members of the genus are nocturnal. Most scorpion stings occur when the arachnids are accidentally touched in their hiding places or when they are prowling in search of food. In such instances, they sting immediately, then beat a hasty retreat. They never display the belligerent, strutting posture assumed by some of the large scorpions of families Scorpionidae and Vejovidae.

Although about 30 species have been listed for the genus, validity of the status of many is in doubt. This situation has resulted because it was not realized that some of the characteristics chosen as criteria for separation of species actually show extensive variation. Two such characteristics, color and presence or absence of an accessory spine on the telson, have been found to be particularly unreliable in this respect. As many as four color phases have been found among examples of a single species. The accessory spine, too, may vary in its presence and in relative size among individuals of a species both in adult and immature stages.

Centruroides sculpturatus Ewing, 1928 [PLATE XVI]

This species, the only truly dangerous scorpion in the United States, has been reported by Ennik (1972) to occur in "western New Mexico, Arizona, in adjacent Mexico and, sporadically, along the west bank of the Colorado River in California." The original description by Ewing was based on two adults and two immature specimens collected in 1927 at Tempe, Arizona.

Ewing described the general color of this species as yellowish brown with no dorsal stripes, spots or other markings. He also write that the integument of both the cephalothorax and abdomen was granular. This led Stahnke (1971) to describe a new species with two dark dorsal stripes which was found in the same general areas as *sculpturatus*. Laboratory studies soon revealed that venom of the new species, *C. gertschi*, was almost identical, both from a biochemical viewpoint and in toxicity, with that of *C. sculpturatus*. Later, Stahnke found that litters produced by patternless scorpions also contained striped individuals. Further study revealed that *C. sculpturatus* was actually represented by four color phases ranging from the characteristic unpatterned straw color to striped specimens much like *C. vittatus* in appearance. Following this discovery, Stahnke (1971) "sunk" the name *Centruroides gertschi* to that of a synonym for *C. sculpturatus*.

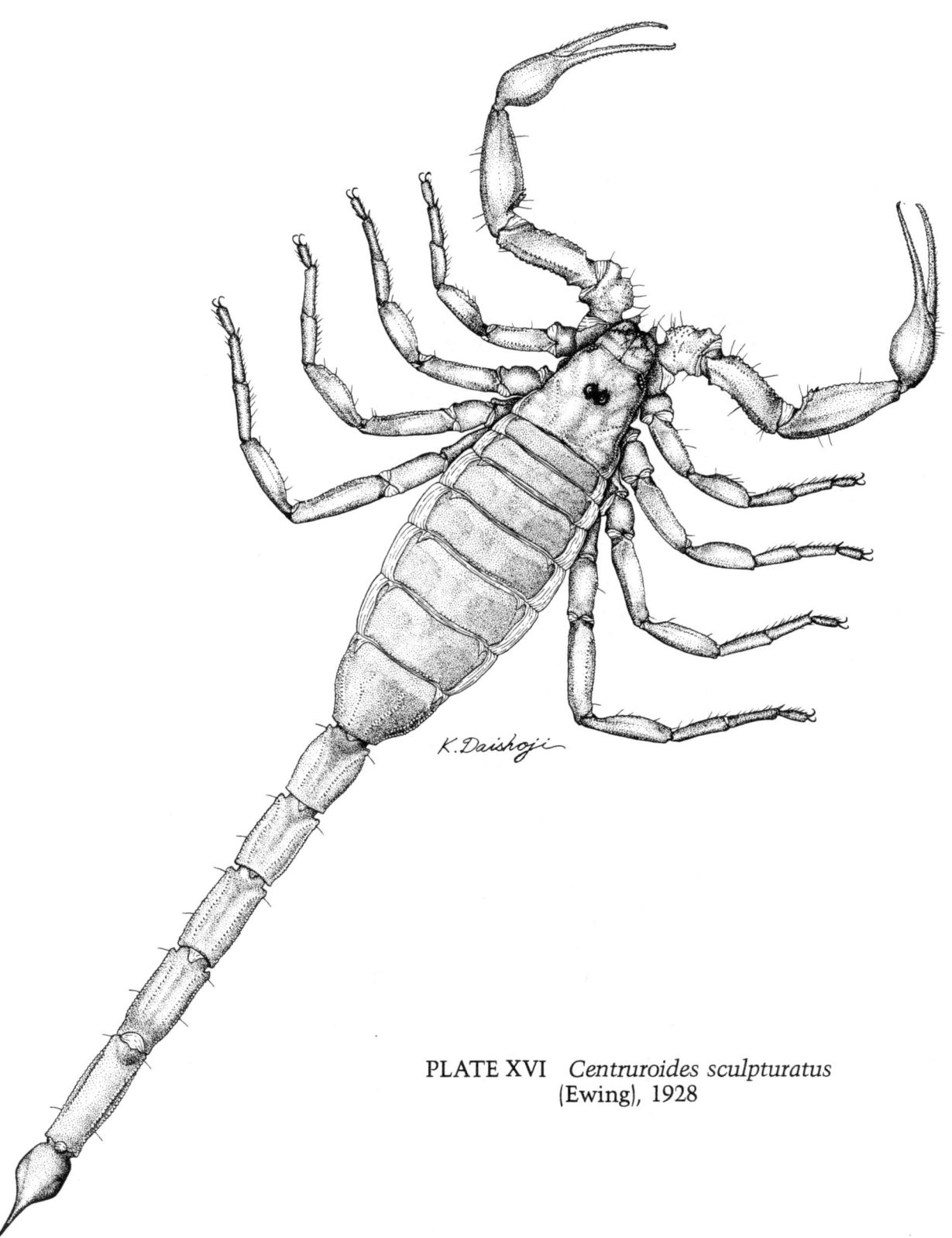

PLATE XVI *Centruroides sculpturatus*
(Ewing), 1928

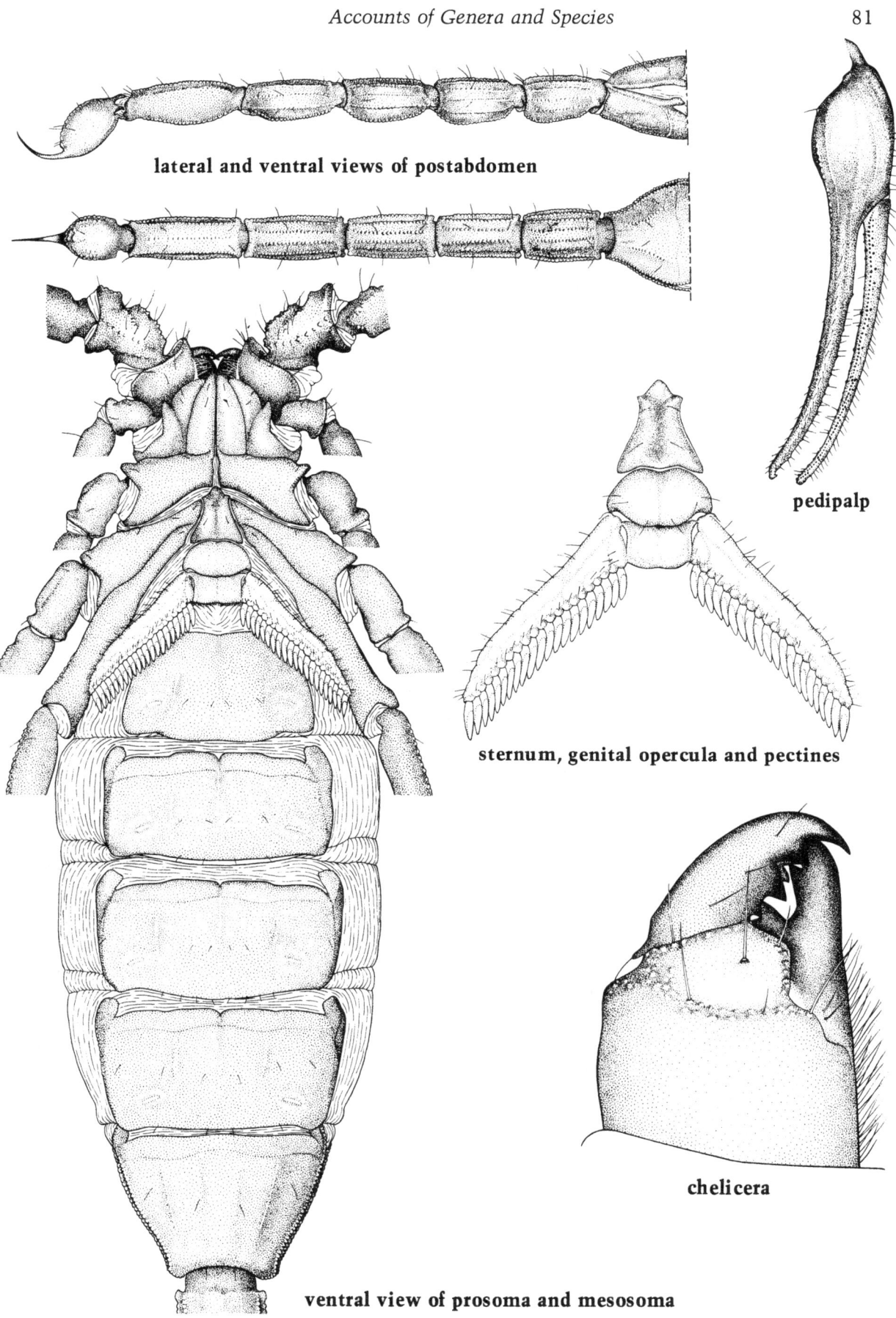

lateral and ventral views of postabdomen

pedipalp

sternum, genital opercula and pectines

chelicera

ventral view of prosoma and mesosoma

Estimations of the duration of the gestation period in *C. sculpturatus* vary considerably according to Stahnke (1966). Production of young by females that had been isolated for more than six months has been reported. Stahnke also noted that the size of the litter has been known to vary from 12 to 36. As with other species of the genus, newly born young ride on the back of the female until shortly after the first ecdysis. In *C. sculpturatus* this occurs six or seven days after birth. Growth is apparently largely dependent upon availability of food. One specimen hatched in captivity survived for five years without reaching maturity.

Details concerning effects of and treatment for the sting of this species are given in chapter 3 of this publication.

Ewing (1928) gave the length of a type specimen as 5.2 cm. In a series of specimens listed by Stahnke (1971) in his re-evaluation of the status of *C. sculpturatus* and *C. gertschi*, length of female plesiotypes varied from 48.47 mm to 55.47 mm. Length of males varied from 53.48 mm to 68.86 mm. The female specimen from Hayden, Arizona illustrated in plate 16 was 57 mm in length. Pectinal teeth of the male ranged in number from 24–29 and those of the female from 19–26 in specimens examined by Stahnke in his 1971 study.

Centruroides vittatus (Say) 1821 [PLATE XVII]

This species, known as the common striped scorpion, is of minor medical importance and is discussed here only because it is the most widely distributed of all American scorpions, and is probably the only species encountered by most persons in the United States. Ewing (1928) wrote that *C. vittatus* had been reported from Georgia, Florida, Kansas, Texas, Arkansas, Louisiana, New Mexico and "doubtless occurs" in all the Gulf States as well as Kentucky and Missouri. Stahnke (1956) described the range of the species as "New Mexico to Florida, including Colorado, Oklahoma, Kansas, Missouri, and southern Illinois." Muma (1967) was of the opinion that the species does not occur in Florida and that records from that state may have been due to mislabeled vials, or that the species may have been introduced but was unable to survive and breed there. *Centruroides vittatus* also occurs in Mexico, where Diáz Nájera (1964) listed records from Tamaulipas, Coahuila, Chihuahua, León and Zacatecas. The species is especially common in Texas, and, except for an occasional specimen of *Vejovis spinigerus*, was the only scorpion encountered by the writer and his colleagues during several years of collecting from San Antonio south to the Mexican border. Like other members of the genus, it is often found under rocks as well as under boards and other debris both outdoors and in houses. In the San Antonio area, specimens could be collected throughout the year in such habitats.

The common striped scorpion is so named because of two broad, brown, sometimes reddish brown, longitudinal stripes on the dorsum of the abdomen. The ground color, also seen on the postabdomen and appendages, varies from yellowish brown to tan in adults. Young specimens may be lighter in color. There is a dark, triangular marking on the anterior portion of the carapace in the area over the median and lateral eyes. This is of particular importance as a means of distinguishing examples of *C. vittatus* from other, sometimes more dangerous, scorpions in areas where ranges overlap. In young specimens, the hands of the pedipalps and the last segment of the postabdomen are dark brown or black.

According to Baerg (1961), *C. vittatus* apparently mates in the fall, and in spring or early summer. He observed mating dances in October and during the first half of July. The gestation period for *C. vittatus* in Arkansas was estimated to be about eight months. Litter size varies from 20 to 47. The average was about 35. Young of the striped scorpion, in examples observed by Baerg, moulted from three to seven days after birth, and remained on the back of the mother for another three to seven days. Whittemore *et al* (1963) maintained adults in the laboratory on a diet of crickets. No attempt was made to rear the young. Smith (1927) estimated the period of growth at three to four years.

The sting of *C. vittatus* causes only temporary local pain and sometimes slight swelling. Although occasional deaths due to scorpion sting in areas where only *C. vittatus* occurs have been reported in newspaper accounts, these have never been substantiated. While it is possible that persons might develop an allergic response following repeated stings, this, too, has never been reported by a reputable source. It is true that dried particles of venom of *C. vittatus*, as well as venoms of other species, are exceedingly irritating to the nasal mucosa. Of course, this is a problem only to persons working with scorpion venoms in the laboratory. Whittemore *et al* (1963) described an improvised device used to draw off airborne venom particles for protection of persons engaged in venom extraction.

Adults of *C. vittatus* probably average around 60 mm in length. The female specimen collected at Camp Bullis, Bexar

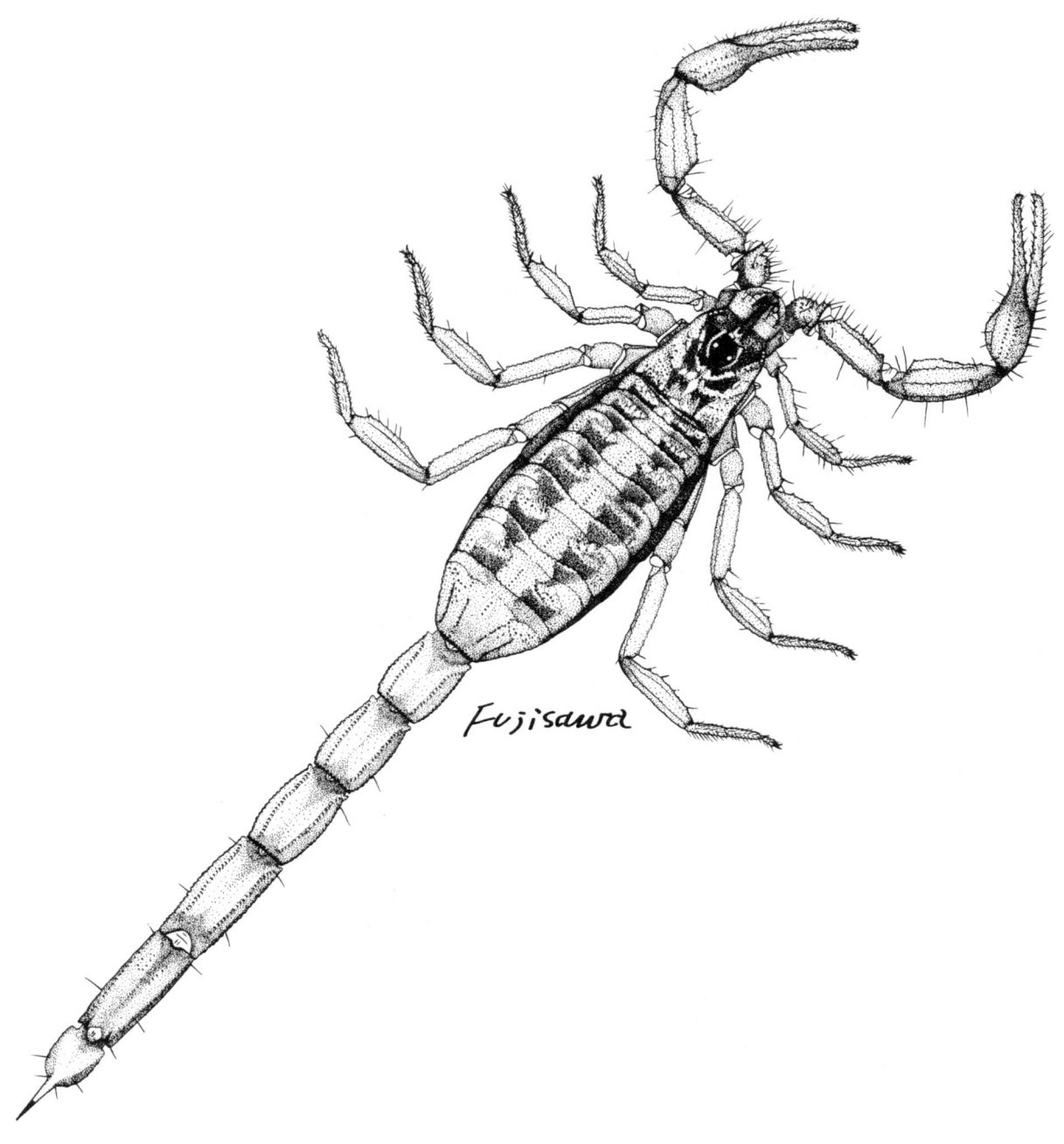

PLATE XVII *Centruroides vittatus*
(Say), 1821

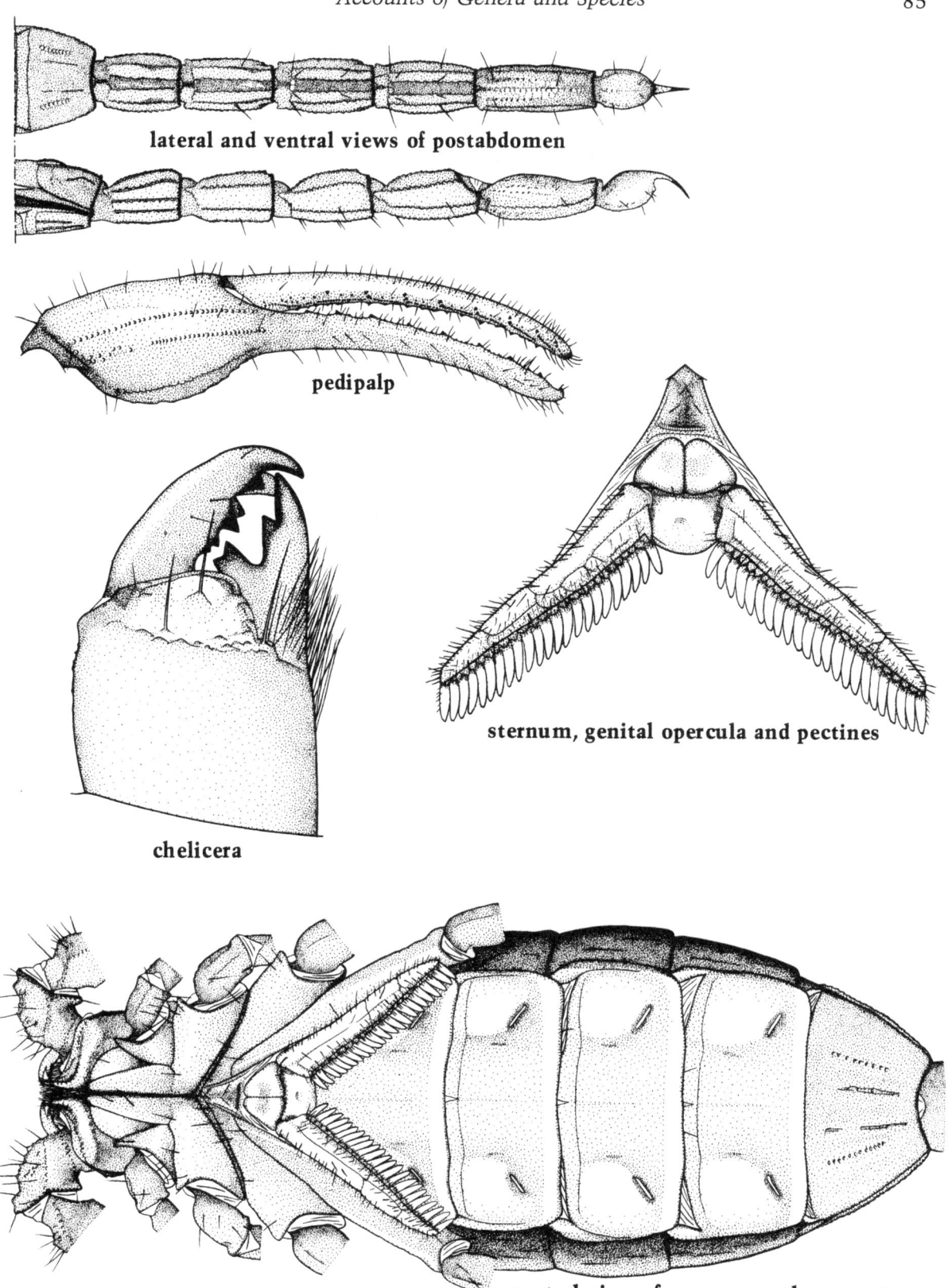

lateral and ventral views of postabdomen

pedipalp

chelicera

sternum, genital opercula and pectines

ventral view of prosoma and mesosoma

County, Texas and figured in plate 17 was 52 mm long. Stahnke (1971) suggested that two light colored species of scorpion from the southwestern United States, *C. pantheriensis* Stahnke and *C. chisosarius,* actually Gertsch represent color pattern phases of *C. vittatus.*

Other Species of Genus *Centruroides* in the United States

None of the remaining species of the genus reported to occur in the United States is of significant medical importance. Three of these, *C. hentzi* (Banks) 1900, *C. keysi* Muma, 1967 and *C. gracilis* (Latreille) 1804, occur in Florida. Muma (1967) has given detailed information on the morphology of each of these and on their distribution in the state. *C. gracilis* has also been reported from Texas, which is not surprising, as the range of the species extends from northern Mexico, through Central America and the Antilles to northern South America. Muma believed that the reported occurrence of *C. margaritatus* (Gervais) 1841 in Florida is doubtful, and may have been due to mislabelling or to misidentification. *C. margaritatus* has a range nearly identical to that of *C. gracilis.* Four additional species have been reported from the western United States. One, *C. nigrescens* (Pocock) 1902, has been reported from the vicinity of San Antonio, Texas and from Eagle Pass, Texas on the Mexican border. In many years of collecting in South Texas, the writer has never encountered this species. It may, however, be present in border areas as it is known to occur in Mexico. *C. exilacauda* (Wood) 1963 has been reported from southern California and in Baja, California. The status of the species may be in question. It was originally thought that absence of a subaculear tooth and the pale yellow color with dark dorsal stripes were valid distinguishing characteristics. However, Stahnke (1971) wrote that instars of the species have a well developed subaculear tooth and that a tubercle may persist in the adult. He also stated that he had data to indicate that *C. exilacauda* has at least three color phases. Similar data led him to question the validity of two additional western species, *C. pantheriensis* Stahnke, 1956 and *C. chisosarius* Gertsch.

This writer has seen no examples of either species and has no opinion on the question.

Mexican Species of *Centruroides*

Of the 28 currently recognized species of *Centruroides* known to occur in Mexico, seven are known to be dangerously venomous. These are: *Centruroides elegans* (Thorell), 1877; *C. infamatus infamatus* (Karsch), 1879; *C. infamatus ornatus*

(Pocock), 1902; *C. limpidus limpidus* (Karsch), 1879; *C. limpidus tecomanus* (Hoffman), 1932; and *C. suffusus suffusus* (Pocock), 1902. An eighth species, *C. sculpturatus* (Ewing) 1928, has been listed by Stahnke and Calos (1977) as occurring in northern Mexico. This is not surprising in view of the distribution of *sculpturatus* in Arizona and New Mexico.

The problem of scorpion sting in Mexico has been reviewed in detail by Mazzotti and Bravo-Becherelle (1963). Although both numbers of accidents and deaths due to scorpion sting have declined in recent years, a total of 20,352 deaths due to this cause were reported in Mexico during the periods 1940–1949 and 1957–1958. States with the greatest scorpion problems recently have been Colima, Guerrero, Nayarit and Morelos. Three-fourths of the deaths reported have occurred among children up to three years of age. While use of antivenin for treatment and insecticides in scorpion control programs have been largely responsible for reduction in fatalities, scorpion sting remains a major public health problem in much of Mexico. An important contributing factor is the poor construction of many houses, particularly in rural areas. While prevention of scorpion infestation is possible through proper construction, including use of screening, both economic and administrative considerations have, to date, prevented significant application of scorpion-proofing techniques.

Centruroides suffusus
Pocock, 1902 [PLATE XVIII]

Until quite recently, any mention of scorpion sting, at least with reference to Mexico, brought the state and city of Durango and *Centruroides suffusus*, the "Alacran de Durango," to the mind of the average student of public health. Actually, as pointed out by Mazzotti and Bravo-Becherelle (1963), during the periods 1940–1949 and 1957–1958 the death rate due to scorpion sting was lower in the state of Durango than in 13 other states of the country. Although Hoffman (1932) and Stahnke and Calos (1977) gave the distribution of *C. suffusus* as Durango only, Diáz Nájera (1964) reported its occurrence in the neighboring state of Zacatecas.

In general appearance, *C. suffusus* is similar to the common striped scorpion of the United States, *C. vittatus*, a species which has been reported, along with *C. suffusus*, from the state of Zacatecas. An important, easily recognizable difference is the lack, in *suffusus*, of the dark interocular triangle marking so prominent in *vittatus*. The color varies from yellowish to light brown and even reddish brown with two dark

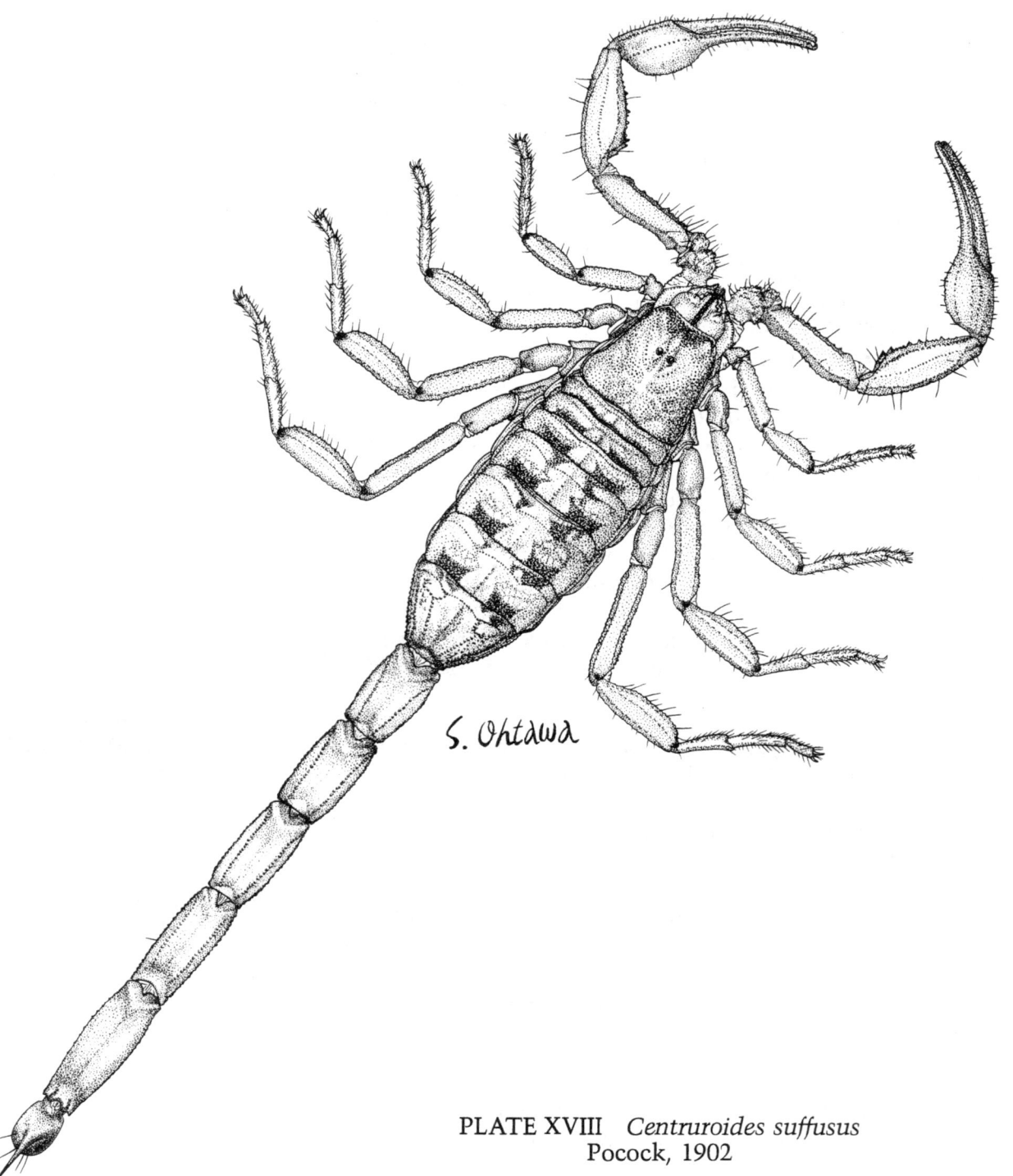

PLATE XVIII *Centruroides suffusus*
Pocock, 1902

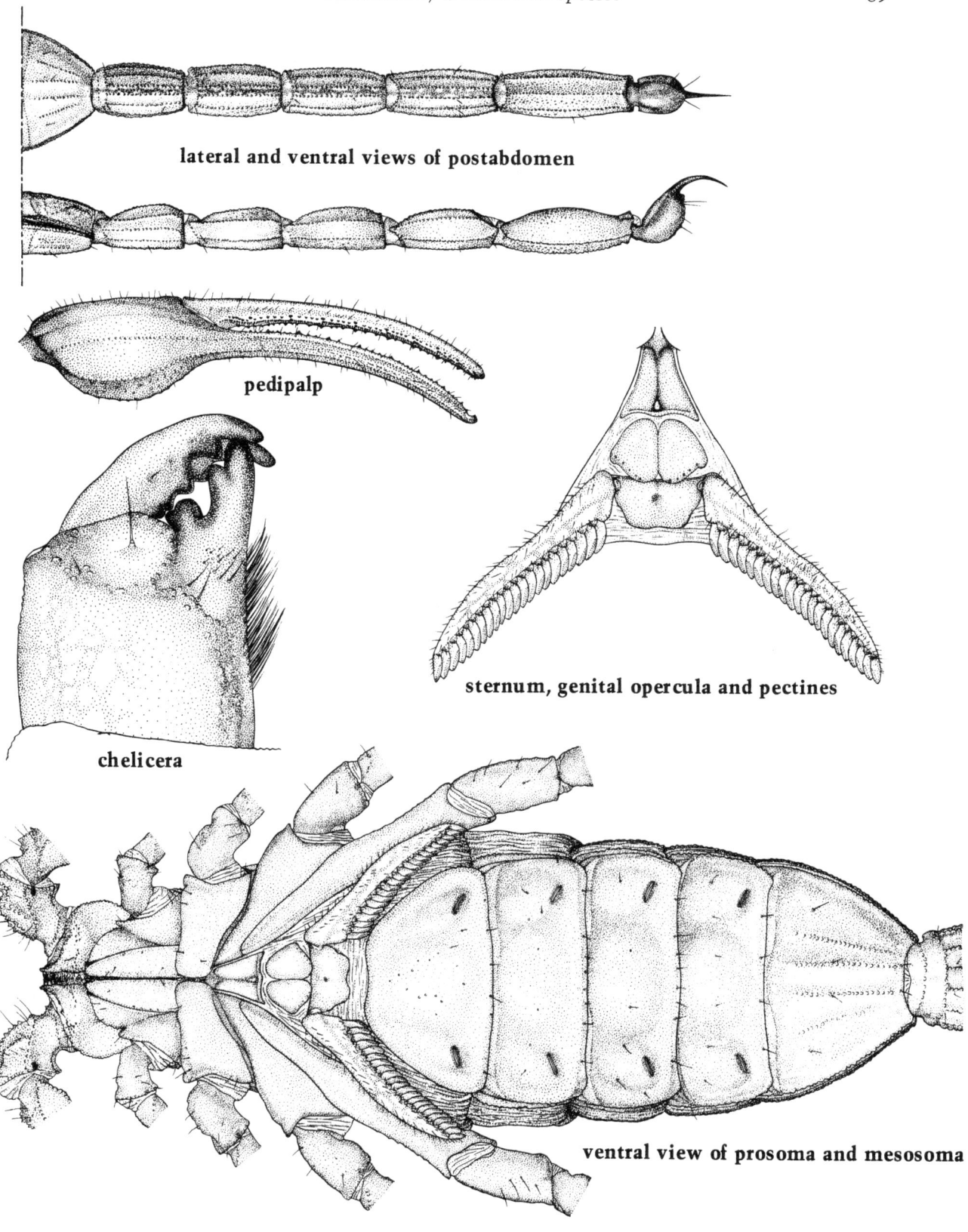

lateral and ventral views of postabdomen

pedipalp

chelicera

sternum, genital opercula and pectines

ventral view of prosoma and mesosoma

longitudinal stripes on the tergites of the preabdomen. In *C. suffusus* males, the length of the postabdomen is at least 8.5 times that of the cephalothorax. In addition, the second caudal segment of the male is slightly longer than the cephalothorax and the fifth caudal segment is 4.25–4.75 times as long as it is wide. In males, the ratio of the length of the vesicle of the telson to the length of the aculeus is over 1.78. These ratios are helpful key characteristics for separation of examples of this species from other Mexican scorpions. Hoffman gave the number of pectinal teeth as 21–26 in males and 20–24 in females, while Stahnke and Calos gave these numbers as 23–25 in males and 20–23 in females. The female specimen illustrated in plate 18 had 22 teeth in either comb. Although *C. infamatus* has also been reported from both Durango and Zacatecas, differentiation of the two species is simple in that the female of *infamatus* lacks the conspicuous central depression or "hole" in the basal piece of the pectines that is present in both *suffusus* and *vittatus*.

While it is hardly necessary to re-document the medical importance of *C. suffusus*, Whittemore and Keegan (1963) found that venom of this species was about equal in toxicity, for white mice, to that of *C. limpidus tecomanus* and about half that of the venom of *C. noxius*. Because of its larger size and venom yield, *C. l. tecomanus* presumably offers a greater hazard to man.

C. suffusus adults attain a length of 8–9 cm. The female specimen from the outskirts of Durango illustrated was 76 mm in length. This specimen along with many others, was collected by SFC J. F. Flanigan on 17 September 1960.

Centruroides noxius
Hoffman, 1932
[PLATE XIX]

This small but extremely venomous scorpion has been reported mainly from the state of Nayarit, although there are a few records from Jalisco and southern Sinaloa. The writer and his colleagues of the U. S. Army Medical Field Service School have collected several hundred specimens from Tepic and San Cayetano in Nayarit.

In this species, the cephalothorax and abdomen are uniformly dark, except that the lateral margins of the tergites of the preabdomen may be lighter in color. This dark color, which may range from black to brown or reddish brown, plus lack of longitudinal stripes on the preabdomen, renders *C. noxious* unique in the areas where it is found. The subaculear tooth is strongly developed in this species. According to

Stahnke and Calos, the pectinal teeth number 17–21 in males and 15–19 in female specimens. The specimen illustrated in plate 19 possesses 23 teeth in either pecten.

This scorpion seldom exceeds a length of 5 cm. The specimen figured in plate 19, was 44 mm in length. It was collected near San Cayetano, Nayarit on 17 September 1960 by the writer.

Centruroides margaritatus
(Gervais), 1841 [Plate XX]

As in the case of *C. vittatus*, this species is discussed here not because of its medical importance. Although its sting causes only transitory pain and discomfort, *C. margaritatus* is noteworthy because it enjoys the greatest geographic range of all the species in genus *Centruroides*. There are well confirmed records of its occurrence from northern Mexico through Central America and the Antilles to South America, where it is commonly found in Venezuela, Colombia, the Guyanas, and Peru.

This scorpion is uniformly dark in color, except that the keels of tergite VII and of the postabdomen, as well as the fifth postabdominal segment are supposed to be conspicuously darker than the remainder of the dorsum. This color difference is not apparent in preserved specimens at hand, which are dark brown to chestnut brown, except that the legs are slighty lighter in color. The vesicle of the telson in the male is said to possess two lateral, terminal expansions on either side of the aculeus. Unfortunately, only female specimens were available for study and no illustrations of this characteristic could be found in other publications. According to Stahnke and Calos (1977), pectinal teeth vary in number from 26–34 in males and 23–32 in females. The female specimen illustrated in plate 20 possesses 28 teeth in each pecten.

After a clinical survey involving over 1000 persons who had been stung by *C. margaritatus* in Colombia, Marinkelle and Stahnke (1965) reported that the venom of this scorpion is of relatively low toxicity for man. The most common signs and symptoms were pain, local edema, and fever, all of which subsided within 24 hours following the sting.

Centruroides margaritatus often exceeds 10 cm in length. The specimen illustrated was 84 mm in length. This specimen, and several other examples, and an excellent drawing in color of a female specimen, were made available through the kindness of Dr. David Botero R. of the School of Medicine, Medillin, Colombia.

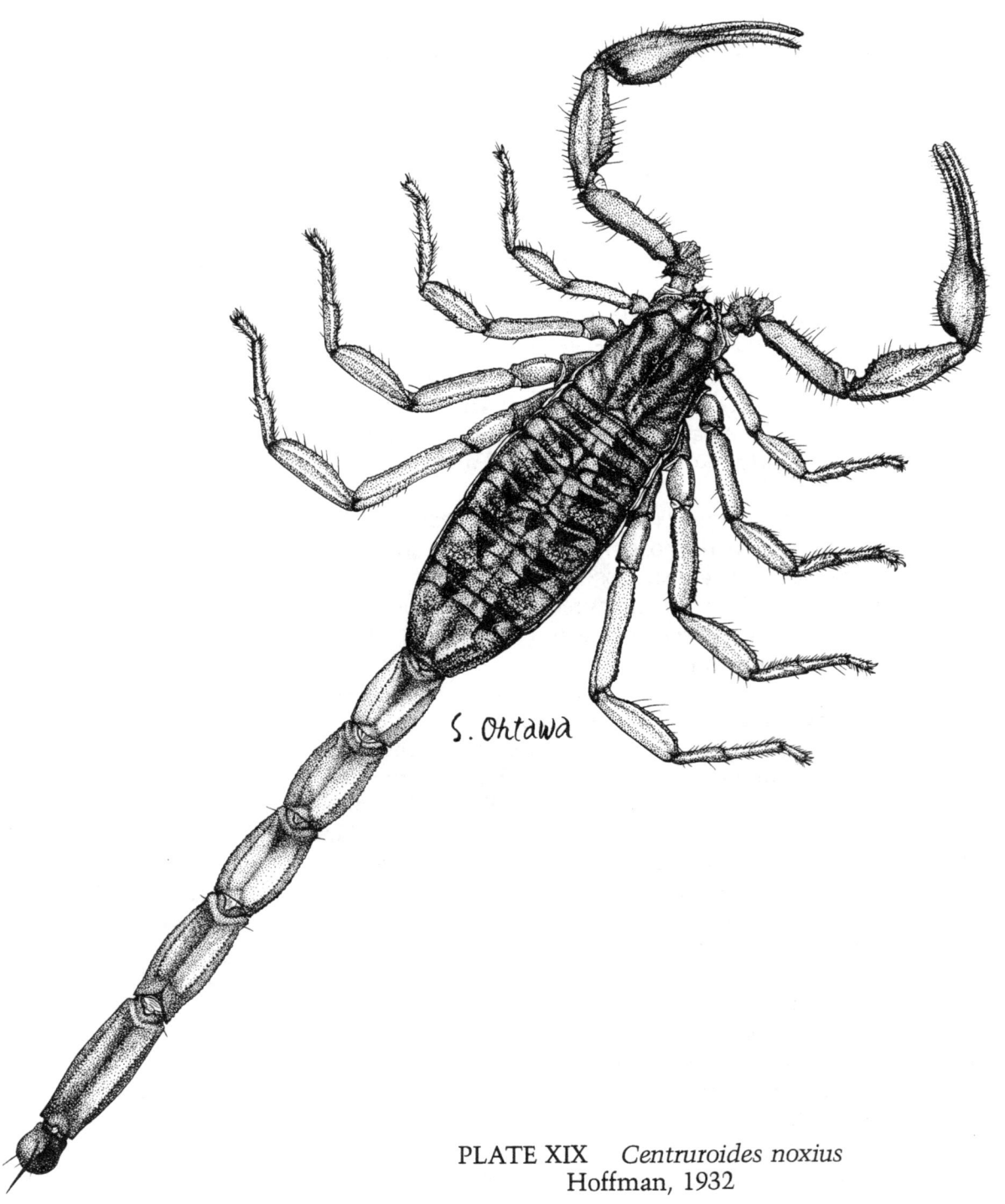

S. Ohtawa

PLATE XIX   *Centruroides noxius*
Hoffman, 1932

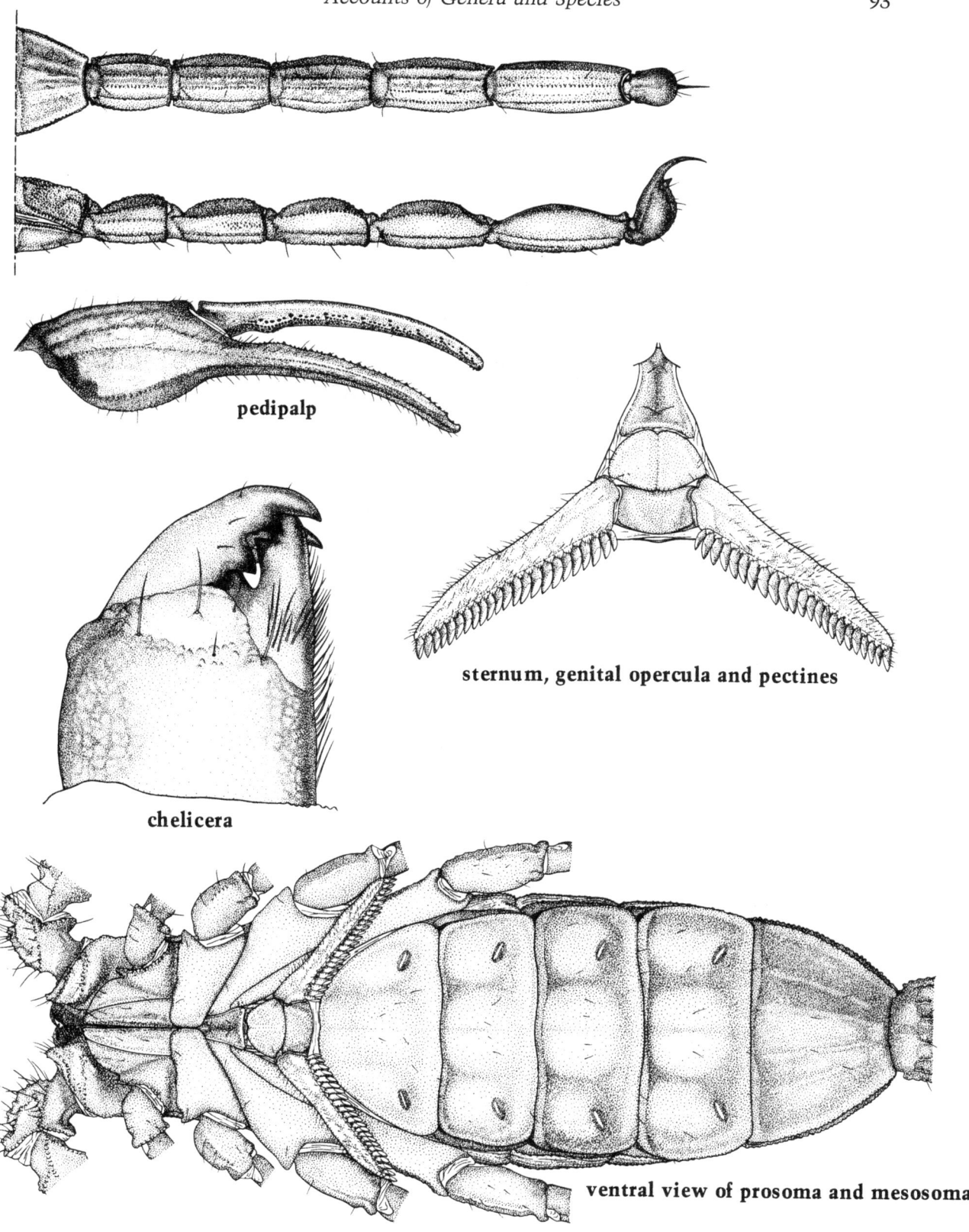

pedipalp

chelicera

sternum, genital opercula and pectines

ventral view of prosoma and mesosoma

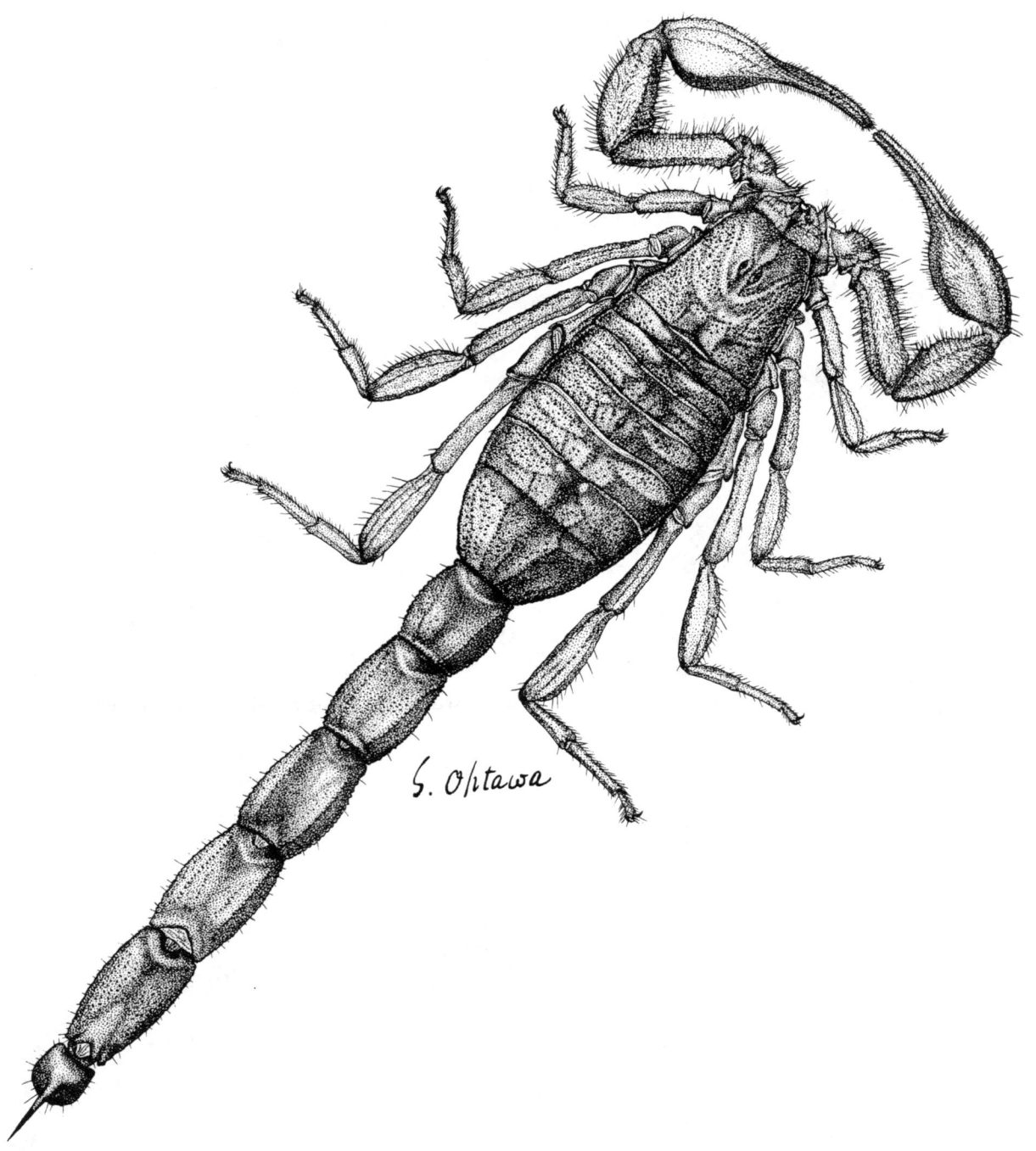

PLATE XX   *Centruroides margaritatus*
(Gervais), 1841

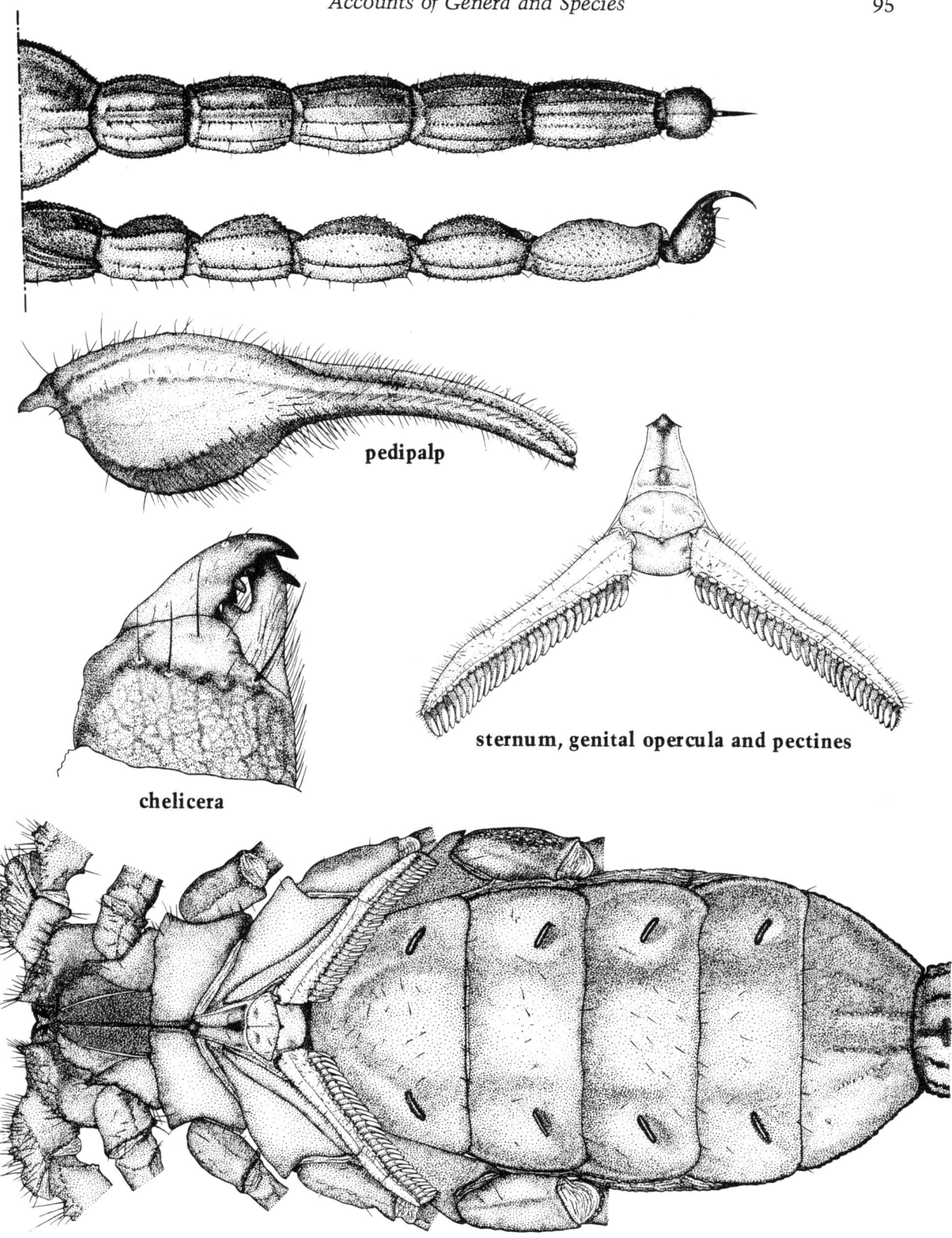

pedipalp

chelicera

sternum, genital opercula and pectines

ventral view of prosoma and mesosoma

Specimens of
Questionable Identity

In their 1977 paper, which included a key to the species of the genus *Centruroides*, Stahnke and Calos pointed out the need for a revisionary study of the genus because earlier descriptions of species had relied too heavily on color and color patterns, characteristics which have been shown to be subject to a high degree of variability. This was clearly illustrated to the writer in his attempt to make specific determinations of several lots of scorpions from various regions in western Mexico. The problem was particularly difficult because of color changes brought about by preservation of these specimens. For this reason, the author is *reasonably certain* of the identity of collections representing only two species: *Centruroides elegans* (Thorell), 1876 and *C. limpidus* (Karsch), 1879. In each case, numerous specimens were available for study. Furthermore, these had been collected in areas close to the type locality for each species.

Centruroides elegans
(Thorell), 1876
[PLATE XXI]

This species has been reported from the states of Guerrero, Jalisco, Nayarit, Sinaloa, Oaxaca and the Tres Marias Islands. Two lots on hand, both from the collection of the United States National Museum, were collected at Iguala, Guerrero and 4 and 5 September 1932 by W. J. Baerg and determined by H. E. Ewing as *C. elegans* (Thor.) var. *limpidus* Karsch.

Centruroides elegans is a member of that group of species of the genus supposedly characterized by the presence of four, well-marked dark lines on the dorsum of the cephalothorax. Scorpions of this species, as in so many of the genus, also possess two longitudinal stripes on the preabdomen. According to Hoffman (1932), a distinguishing characteristic is the presence of an intense black spot on the anterior margin of each preabdominal tergite (in each of the longitudinal stripes) and absence of a corresponding spot on the posterior margin of the tergite. He noted that in some specimens, particularly large females, there were some diffused black spots which united with the anterior spots but did not extend to the posterior margin of the tergite. In the preserved material at hand, the stripes on the cephalothorax have almost completely faded, and the markings on the preabdominal tergites are not clear. One characteristic not mentioned by Hoffman, but present in all examples from Iguala, is the presence of a centrally located depression in the basal piece of the pectines. Because of the condition of these specimens, the writer must admit that he is classifying these as examples of *C. elegans* primarily because of the determination made by H. E. Ewing who, at that time, was the foremost authority of scorpions in the United States.

Adult males are said to reach a length of 8 cm, the females slightly less. The female specimen shown on plate 21 was 68 mm in length.

Hoffman (1932) considered this species to be properly divided into two subspecies, *C. limpidus limpidus* (Karsch), 1879 and *C. limpidus tecomanus* Hoffman, 1932. Distribution of *C. l. limpidus* was given as the central region of the state of Guerrero, the state of Morelos and the southern part of the state of Pueblo, while *C. l. tecomanus* was said to occur only in Colima. Later Díaz Nájera (1964) reported collections of *C. l. tecomanus* from Colima.

As in the case of *C. elegans*, this species belongs to the group in which the four dark lines on the cephalothorax are supposedly well-defined in both adult and immature specimens. On each preabdominal tergite (in the series of markings which make up the longitudinal stripes), is a dark spot (one on either side of the clear mid-dorsal area) on the anterior margin of the tergite and another on the posterior margin. Both spots are diffusely united. Hoffman considered that division of the species was justified because the subaculear tooth was well developed in adults of *C l. tecomanus* and reduced to a small tubercle in *C. l. limpidus*.

The writer has collected large numbers of *C. limpidus* from Colima City and Manzanillo, Colima and with the exception of some differences in the shape of the genital opercula and the basal piece of the pectines can find no differences between the two. In each group the subaculear tooth (spine) is greatly reduced. In each also, the fingers of the pedipalps are distinctly darker than the hand and the other segments of the pedipalp. Due to fading caused by preservation, the dorsal stripes on the cephalothorax are not visible. The typical marking described by Hoffman is best seen in the specimen from Colima illustrated on plate 22.

Centruroides limpidus is a species of considerable medical importance. A dramatic example of this was afforded in October 1959 when a hurricane struck the coastal city of Manzanillo causing widespread property destruction and knocking out utilities. To add to the problems of the inhabitants, swarms of scorpions, driven from their hiding places in flood-crumpled adobe walls, stung several hundred persons. During this crisis, locally available scorpion antivenin

Centruroides limpidus (Karsch), 1879 [PLATE XXII]

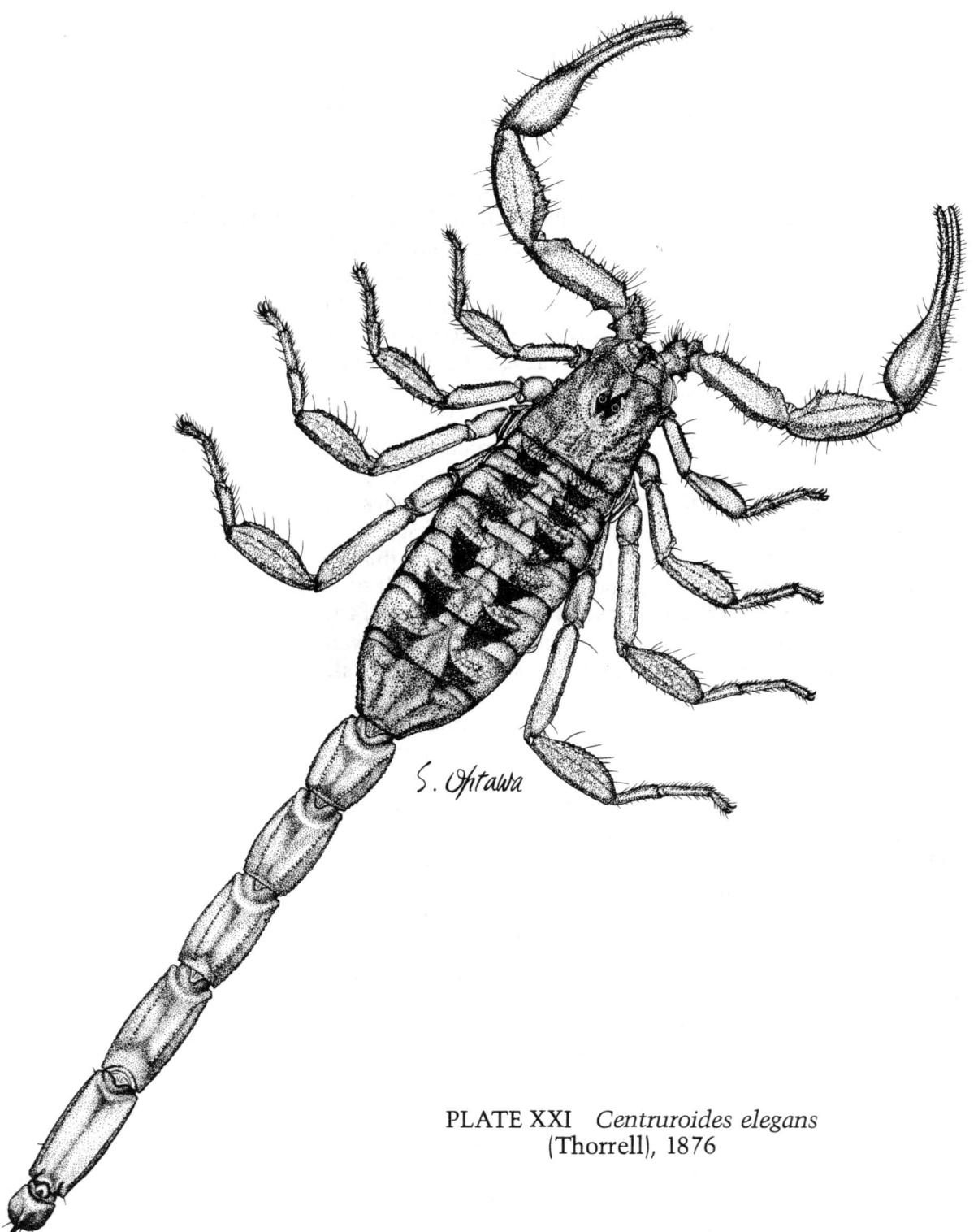

PLATE XXI   *Centruroides elegans*
(Thorrell), 1876

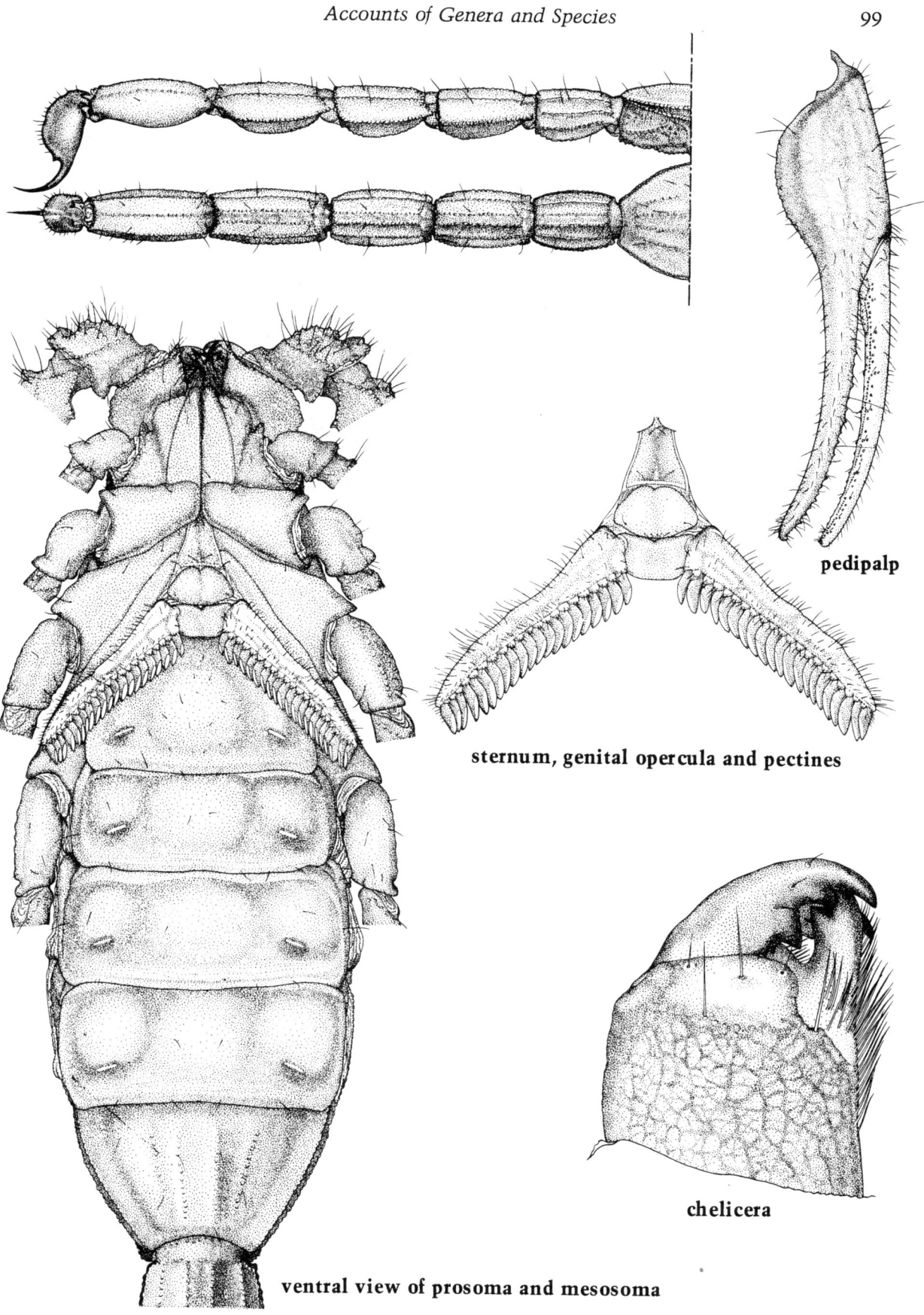

pedipalp

sternum, genital opercula and pectines

chelicera

ventral view of prosoma and mesosoma

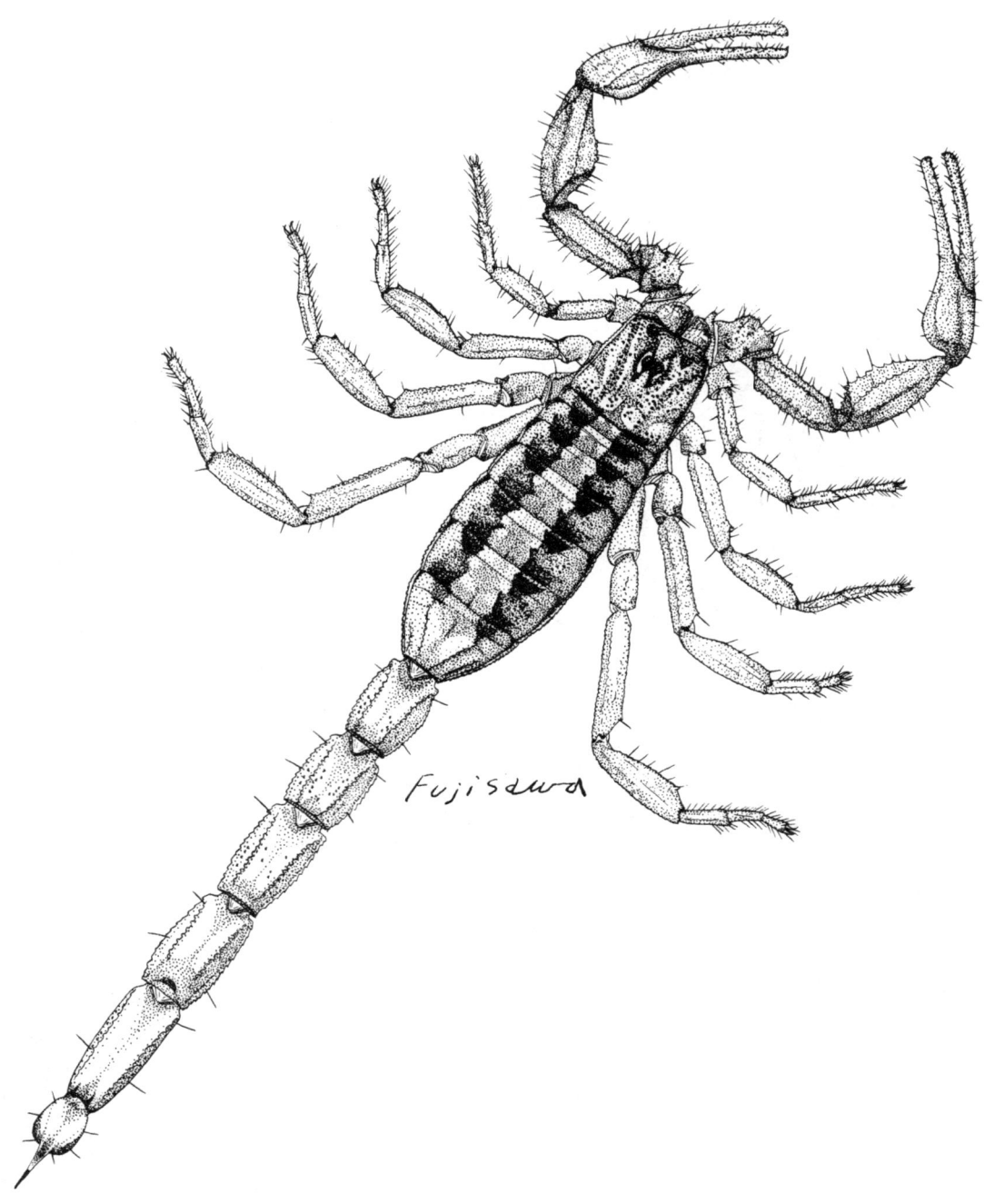

PLATE XXII *Centruroides limpidus*
(Karsch), 1879

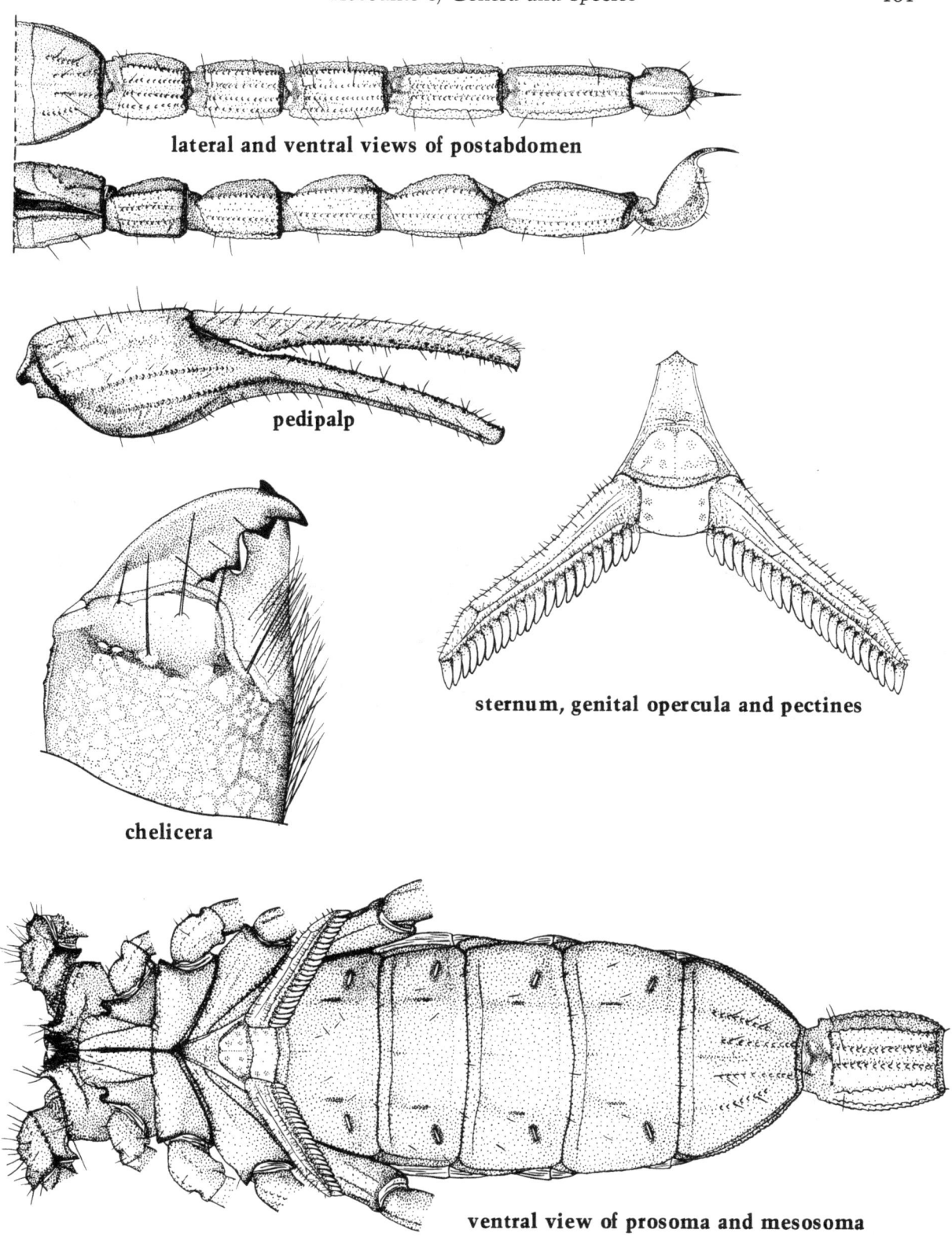

lateral and ventral views of postabdomen

pedipalp

chelicera

sternum, genital opercula and pectines

ventral view of prosoma and mesosoma

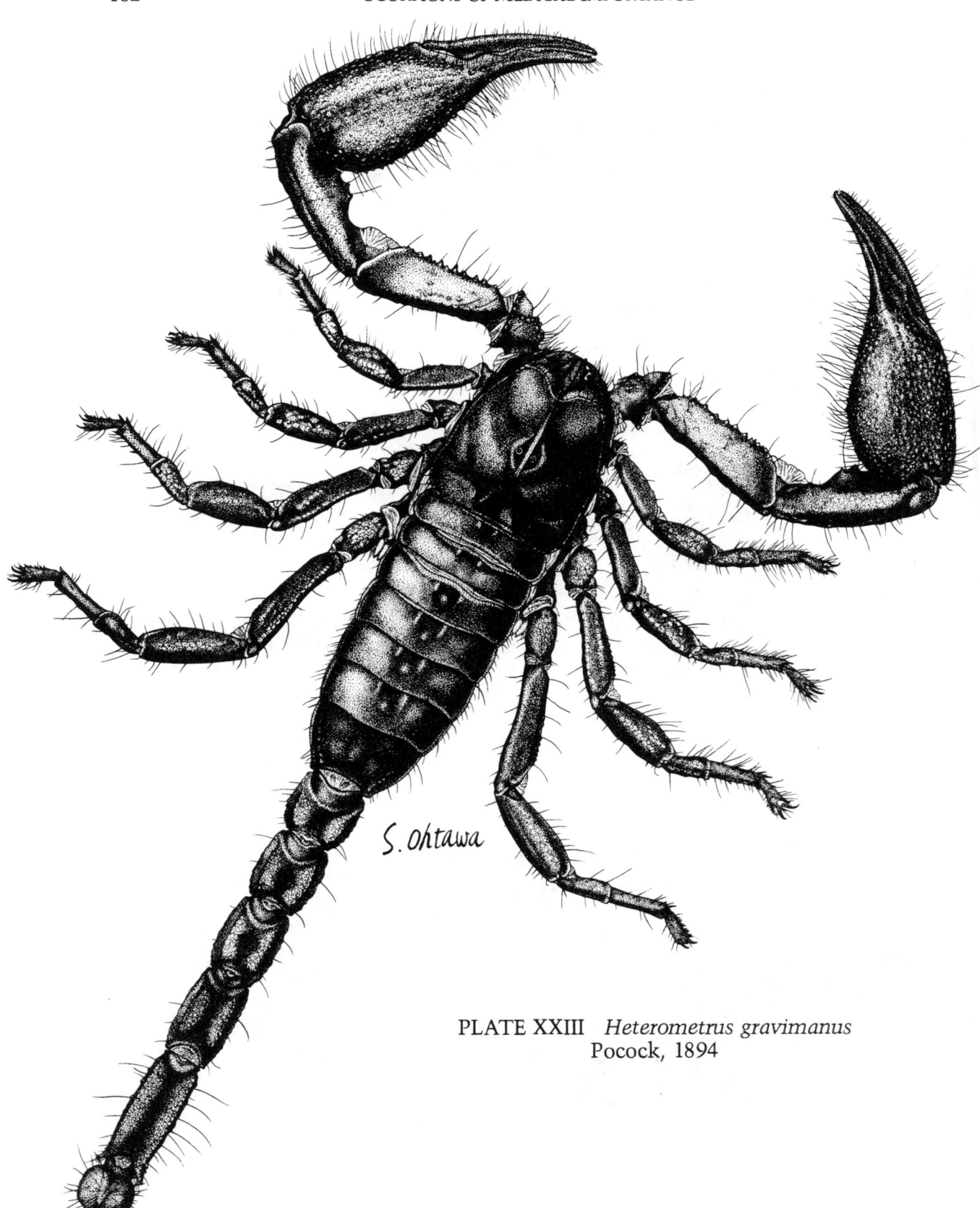

PLATE XXIII   *Heterometrus gravimanus*
Pocock, 1894

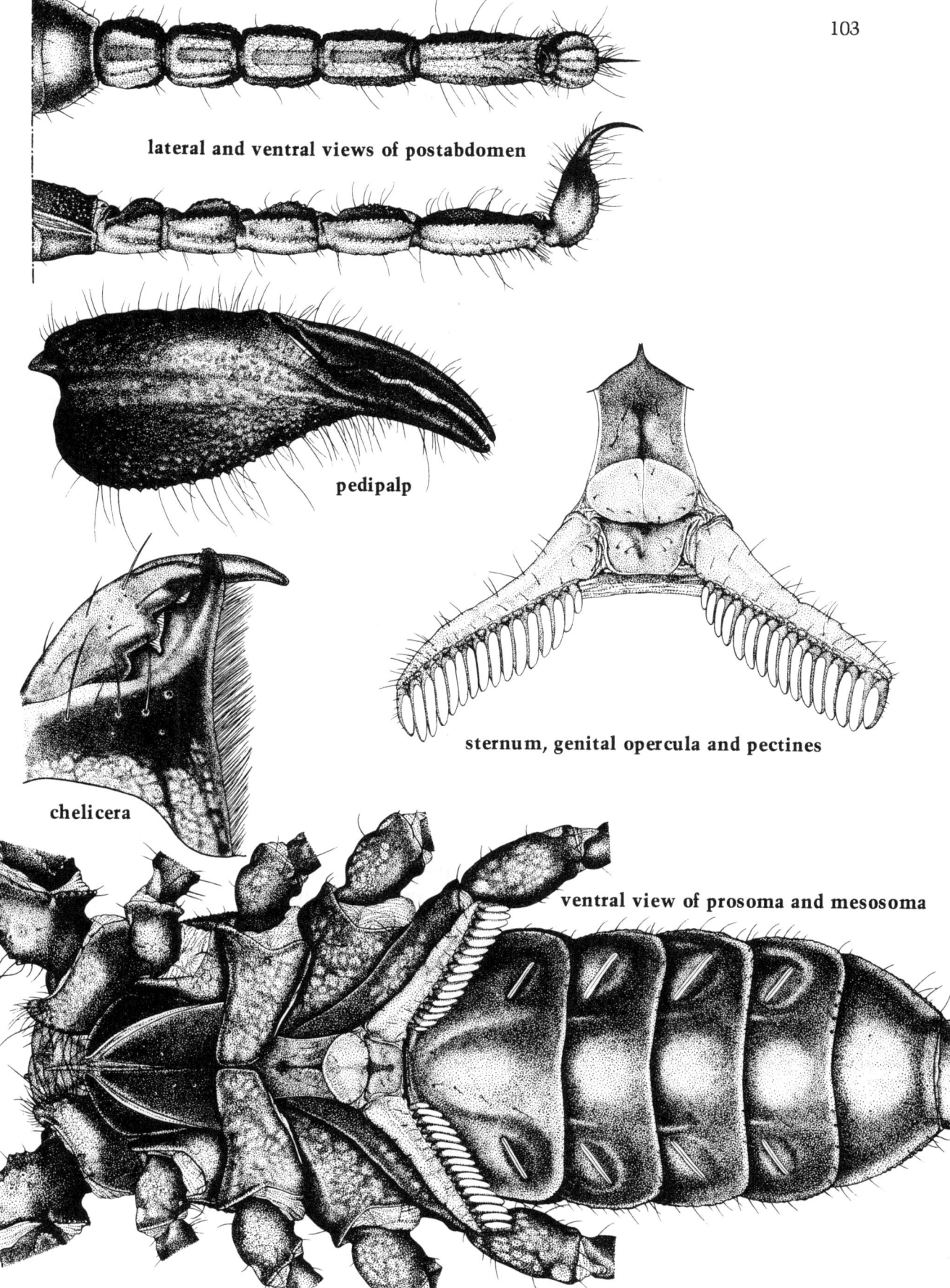

lateral and ventral views of postabdomen

pedipalp

sternum, genital opercula and pectines

chelicera

ventral view of prosoma and mesosoma

supplies were exhausted and difficulty in transportation left many victims in outlying areas without medical assistance.

In tests with white mice, Whittemore and Keegan (1963) found venom of *C. limpidus* to be approximately equal to that of *C. suffusus* in virulence. Oddly, Keegan and Lockwood (1971) working with specimens from the city of Colima, could find no morphological differences in secretory epithelium of venom glands of *C. limpidus* and the relatively innocuous *C. vittatus* in studies involving both light and electron microscopy.

The species is said to attain a length of about 7 cm. The female specimen illustrated on plate 22 collected in Manzanillo, Colima by the author on 6 January 1961, was 63 mm in length. The specimen illustrated on plate 22, collected in the city of Colima by J. Sanchez Garcia in December 1970, was 58 mm in length.

FAMILY SCORPIONIDAE

Genus *Heterometrus*
Hempr. & Ehrenb., 1828

Their large size, plus the deeply indented anterior margin of the carapace and the broad hand of the pedipalp (plate 23), serves to distinguish species of this genus from other scorpions in areas where they occur. Species of *Heterometrus*, along with the African species of *Pandinus*, are the world's largest scorpions. Some attain a length of 180 mm or more. There are probably about 15 valid species. These are widely distributed from Central to Southeast Asia. Keys to the species have been published by Kraepelin (1899) and Pocock (1900 a. and b.). Although species of the genera *Pandinus* and *Heterometrus* do not occur in the same geographic areas, they show remarkable similarity. A distinguishing characteristic of *Heterometrus* is the presence of only two spines on each of the lateral lobes of the terminal tarsal segment. In scorpions of genus *Pandinus* there are frequently four spines on each of these lobes. Species of both *Heterometrus* and *Pandinus* possess a stridulatory apparatus consisting of hairs and tiny spurs on opposing portions of the maxillary process of the coxa of leg 1 and on the pedipalp. It is said that a hissing sound is produced when the two surfaces are rubbed together. However, during observation of several hundred specimens of *Heterometrus gravimanus* caught in the wild and over a hundred laboratory reared adults of the same species, this habit was never observed by the author. Specimens on hand for study and illustration included examples of *H. longimanus* (Herbst), *H.*

cyaneus (Koch), *H. gravimanus* Pocock, and H. *silenus* (Simon).

Although Pocock described *H. gravimanus* as "reddish black, legs dark reddish brown," several hundred living specimens maintained in the laboratory by the writer mainly fulfilled the black rather than the "reddish" aspect of the description. Although these scorpions were quite belligerent and would spread their pedipalps and arch their tails at the slightest provocation, a sting from even the largest specimen would seldom kill a mouse. Deoras (1961) and Whittemore *et al* (1963) found that *H. gravimanus* was easily kept in the laboratory on a diet of roaches or crickets. Studies of the chromosomes of this species were made by Yoshida and Toshioka (1964). Deoras reported that BHC in water dispersible suspensions or powder formulations was lethal for these scorpions. Conditions under which these tests were carried out were not described.

Adults in our laboratory colony varied in length from 90–150 mm. The specimen shown on plate 23 was 117 mm long. Pocock gave the geographic distribution of the species as south India and Ceylon. Our specimens were obtained from a dealer in Bombay so their exact origin is unknown.

Heterometrus gravimanus Pocock, 1894 [PLATE XXIII]

The color of this species, as described by Pocock (1900) is "a uniform black or deep blackish brown, reddish brown below and on the hands and vesicle." Kraepelin (1899) wrote that it was "dark chestnut brown or dark green to black with a yellowish vesicle." Schultze (1927) described adults in his collection as black. Preserved specimens on hand are uniformly dark brown. It is likely, of course, that colors of the living specimens were lost in preservation.

Schultze (1927) who described *H. longimanus* as the Philippine forest scorpion, wrote that specimens were usually found in old or virgin forests under the loose bark of dead standing trees, under decaying trunks of trees and logs, or in cavities of rotten stumps located in the jungle, mostly in rather humid and damp places. Specimens kept by him in the laboratory took cockroaches in preference to other insects offered as food. He noted that the sting was never used to subdue the prey. A female specimen from Mindanao in his possession gave birth to 33 young within a 24-hour period. These young scorpions moulted for the first time eight days following birth. Within three days after the first moult, more

Heterometrus longimanus (Herbst), 1800 [PLATE XXIV]

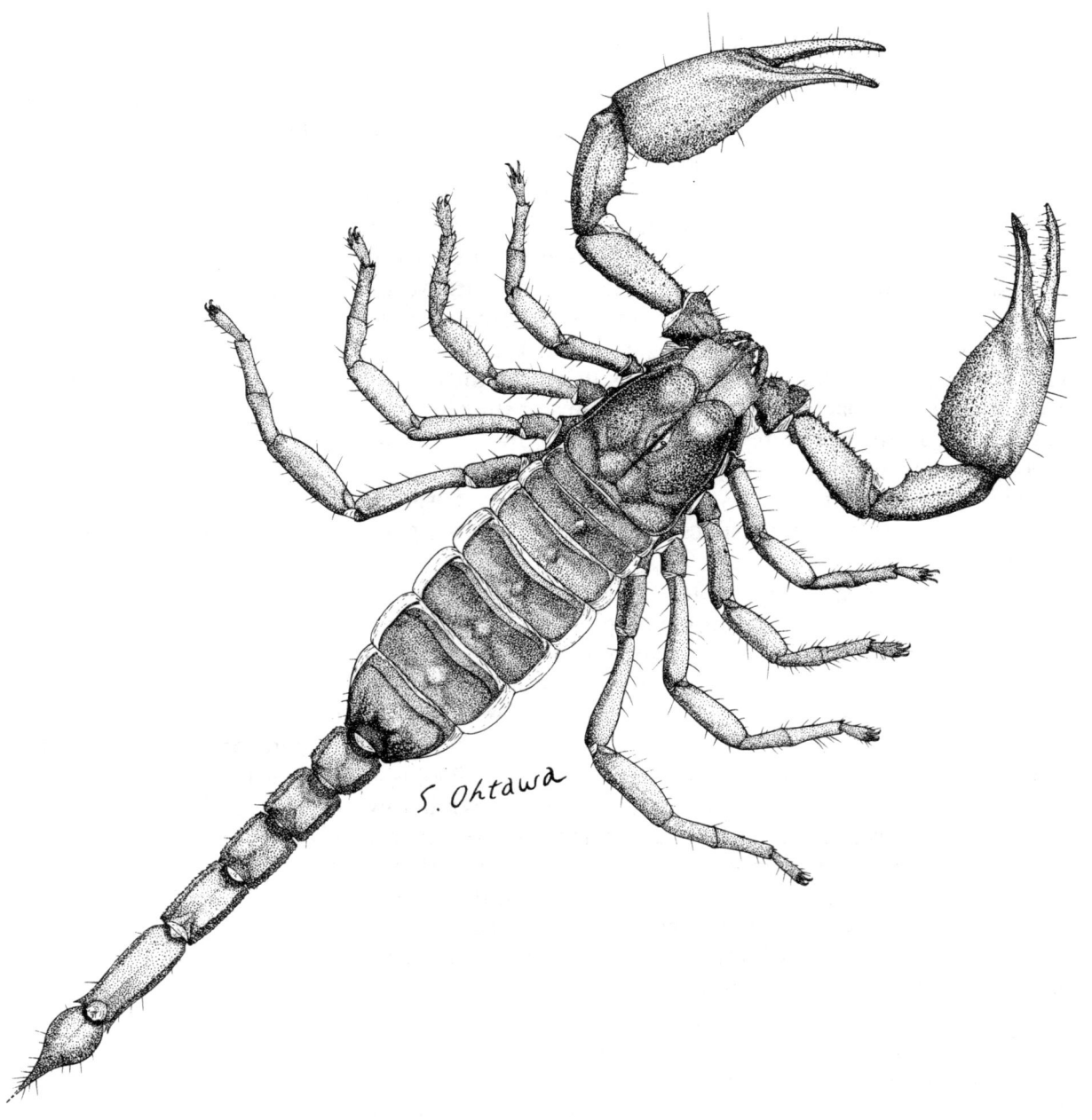

PLATE XXIV   *Heterometrus longimanus*
(Herbst), 1800

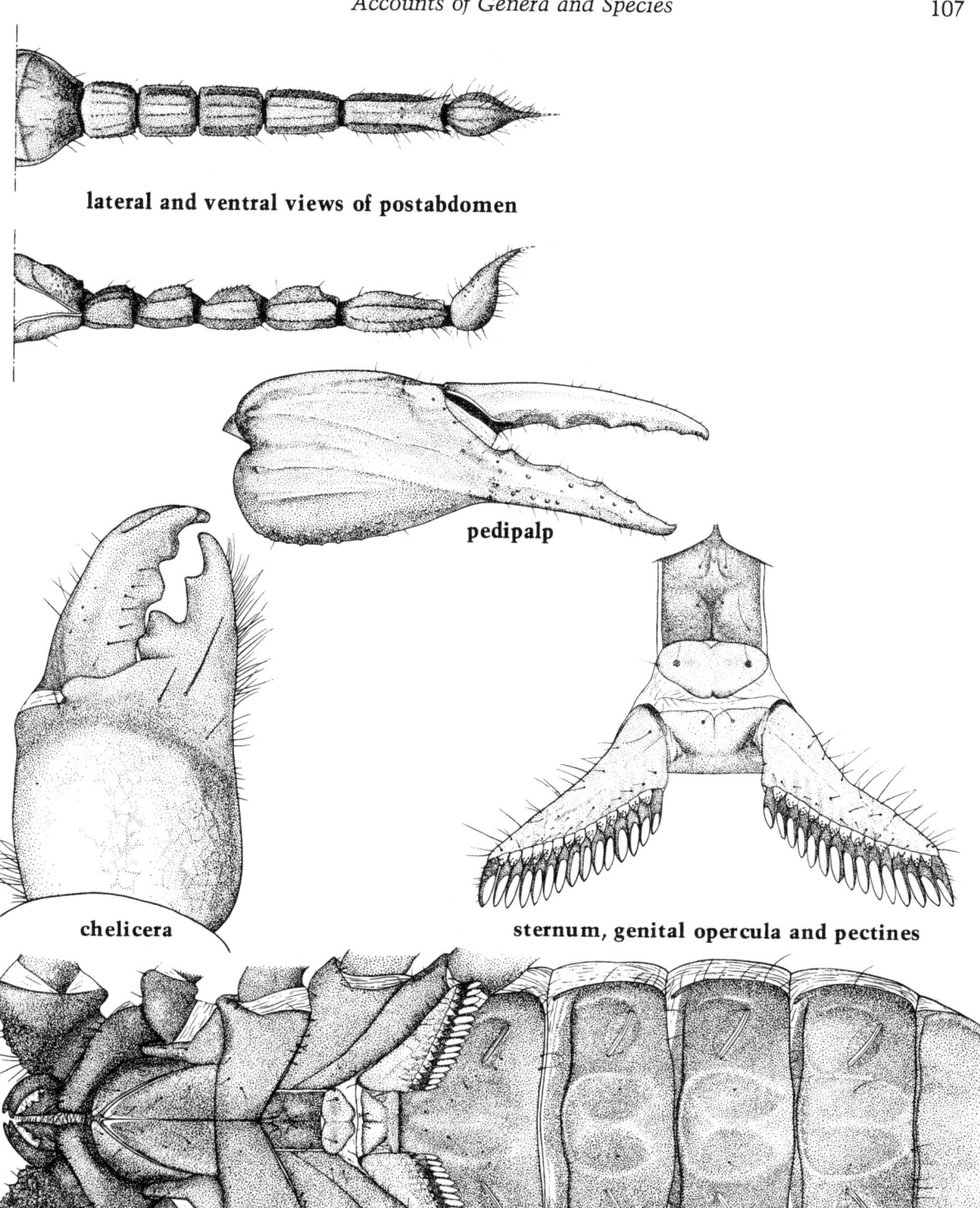

lateral and ventral views of postabdomen

pedipalp

chelicera

sternum, genital opercula and pectines

ventral view of prosoma and mesosoma

than 50% of the young had left the back of the mother, but were located close to her on the substrate. Subsequent moults, as with other species, seemed to be dependent upon food supply and were quite irregular. One specimen reached the adult stage in 345 days, and eight became adults in 403–464 days following birth. At that time, four specimens had not yet undergone the seventh moult. Schultze believed the life span of this species to be not less than two years, and possibly three years.

The only published account of effects of a sting by *H. longimanus* which the writer has been able to find is that by Kopstein (1932). In this instance, a field worker stung on the toe experienced pain and swelling of the leg up to the groin. By evening of the same day, the symptoms had subsided and the man was able to return to work the next day.

H. longimanus has been found from Burma eastward throughout Southeast Asia. There are specimens in the United States National Museum from Sumatra, Thailand, Malaysia, Borneo, and the Philippines.

According to Kraepelin, adults of this species are usually from 100–110 mm in length. The specimen from Luzon, Republic of the Philippines shown on plate 24 was 98 mm long.

Heterometrus cyaneus
(Koch), 1836
[PLATE XXV]

Kraepelin described the color of this species as chestnut brown to dark brown or dark green; the legs, toward the ends, as yellow to reddish brown.

Although examples of this species have been known to reach a length of 140 mm, effects of their stings have been transitory and not severe. Envenomation by this scorpion was discussed by Kopstein (1927 and 1932). Severe pain immediately followed the sting and considerable swelling was also present. However, these symptoms were of short duration. In laboratory tests it was found that stings by adult scorpions were fatal to small birds but not to chickens, guinea pigs, and rats.

There are numerous specimens of *H. cyaneus* in the United States National Museum collections. All of these are from Java. The length of the specimen figured was 125 mm.

Heterometrus silenus
(Simon), 1872
[PLATE XXVI]

Nothing is known concerning the venom of this species. Kraepelin wrote that the color was deep black. Distribution of the species was given as Cochin China (South Vietnam). There are specimens in the United States National Museum from Cochin China and from Thailand. The specimen illustrated was collected at the Trang Bom Arboretum at Bienhoa near Saigon in 1932. This example was 130 mm in length. Kraepelin considered *H. silenus* to be only a subspecies of *H. longimanus*.

Species of the genus are widely distributed in tropical Africa and on the Arabian peninsula. As in genus *Heterometrus*, the lateral lobes of tarsomere II have rounded terminal margins. In *Pandinus* the hand of the pedipalp lacks finger keels, but may have keel-like swellings formed by fused rows of granules. The anterior margin of the carapace has a deep median indentation. Vachon (1973) has divided genus *Pandinus* into 5 subgenera and 23 species, mainly on the basis of the number and arrangement patterns of the trichobothriae. Vachon (1966) and Probst (1973) have discussed distribution of various species of the genus. Balozet (1971) mentioned the relatively innocuous nature of the venom of these large scorpions. There had been no published evidence of other than local, transitory pain from stings by species of *Pandinus*. In spite of this, the large size (up to 180 mm or more), plus the belligerent behavior of these arachnids when they are disturbed, has made them greatly feared in many areas. Observations of Vachon *et al* (1970) on the post-embryonic development of one species, *P. gambiensis*, have been outlined in chapter 1 of this publication. Examples of only two species, *P. imperator* C. L. Koch, 1842, and *P. dictator* (Pocock, 1888) were on hand for illustrations.

This, perhaps the largest of the scorpions, has been described as varying in general coloration from dark green to earth brown, with the telson and pedipalps sometimes reddish brown. The preserved specimens on hand were uniformly dark brown. Kraepelin (1899) gave the distribution of the species as West Africa (Congo to Gaboon) and listed a subspecies, *P. imperator subtypicus* (Kraepelin, 1875) from Sudan. The latter was not mentioned by Vachon (1973) in his review of the genus. The specimen illustrated was collected at Bandaja, Liberia by W. M. Mann during the Firestone Expedition in 1940. This scorpion was 152 mm in length.

Until Vachon (1973) erected the subgenus *Pandinopsis*, *dictator* was regarded as a synonym for or, at most, a subspecies of *imperator*. No information is available concerning habits of

Genus *Pandinus* Thorell, 1877 (em. Krpln.)

Pandinus imperator
C. L. Koch, 1842
[PLATE XXVII]

Pandinus (Pandinopsis)
dictator (Pocock, 1888)
[PLATE XXVIII]

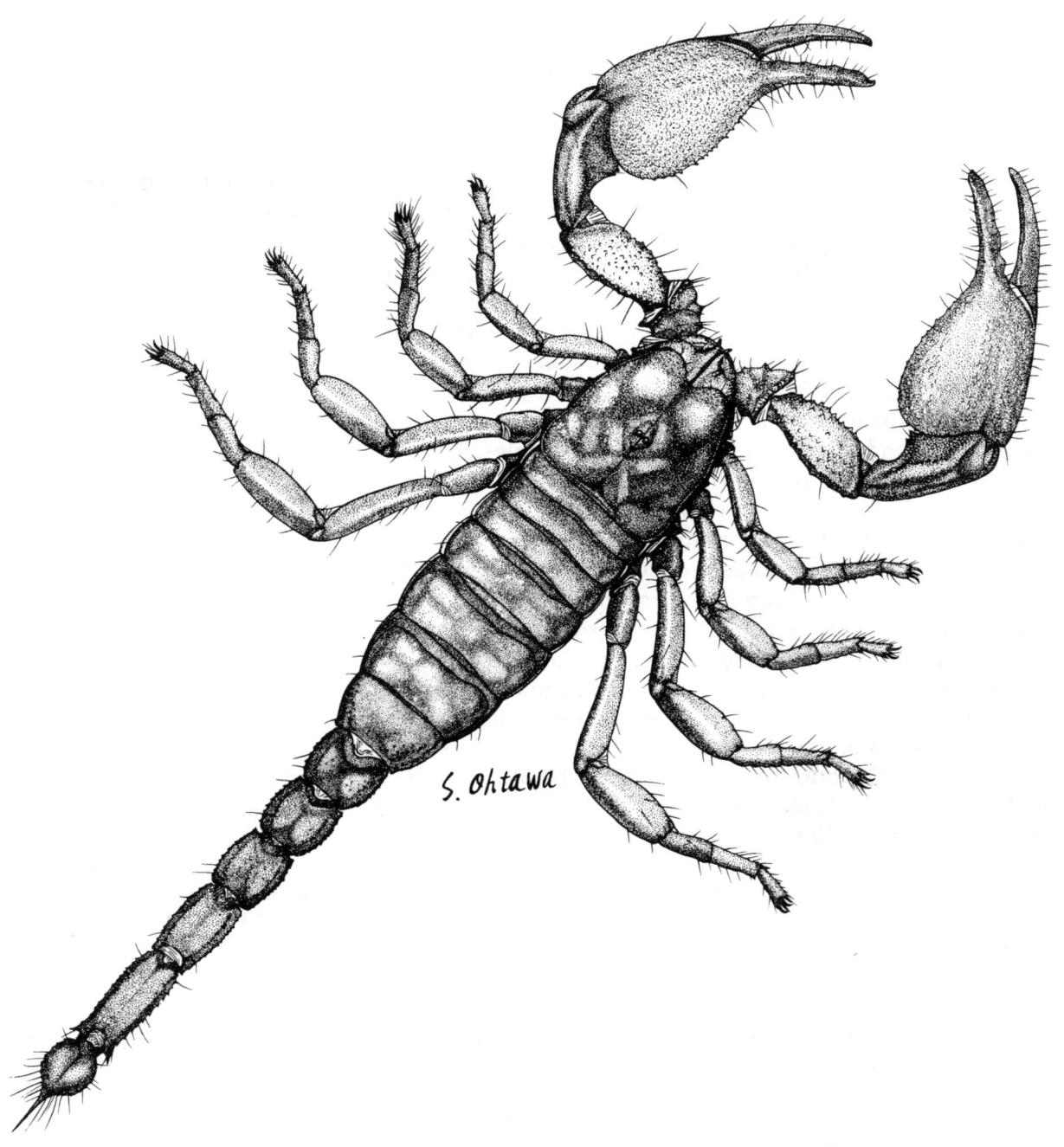

S. Ohtawa

PLATE XXV *Heterometrus cynaeus*
(Koch), 1836

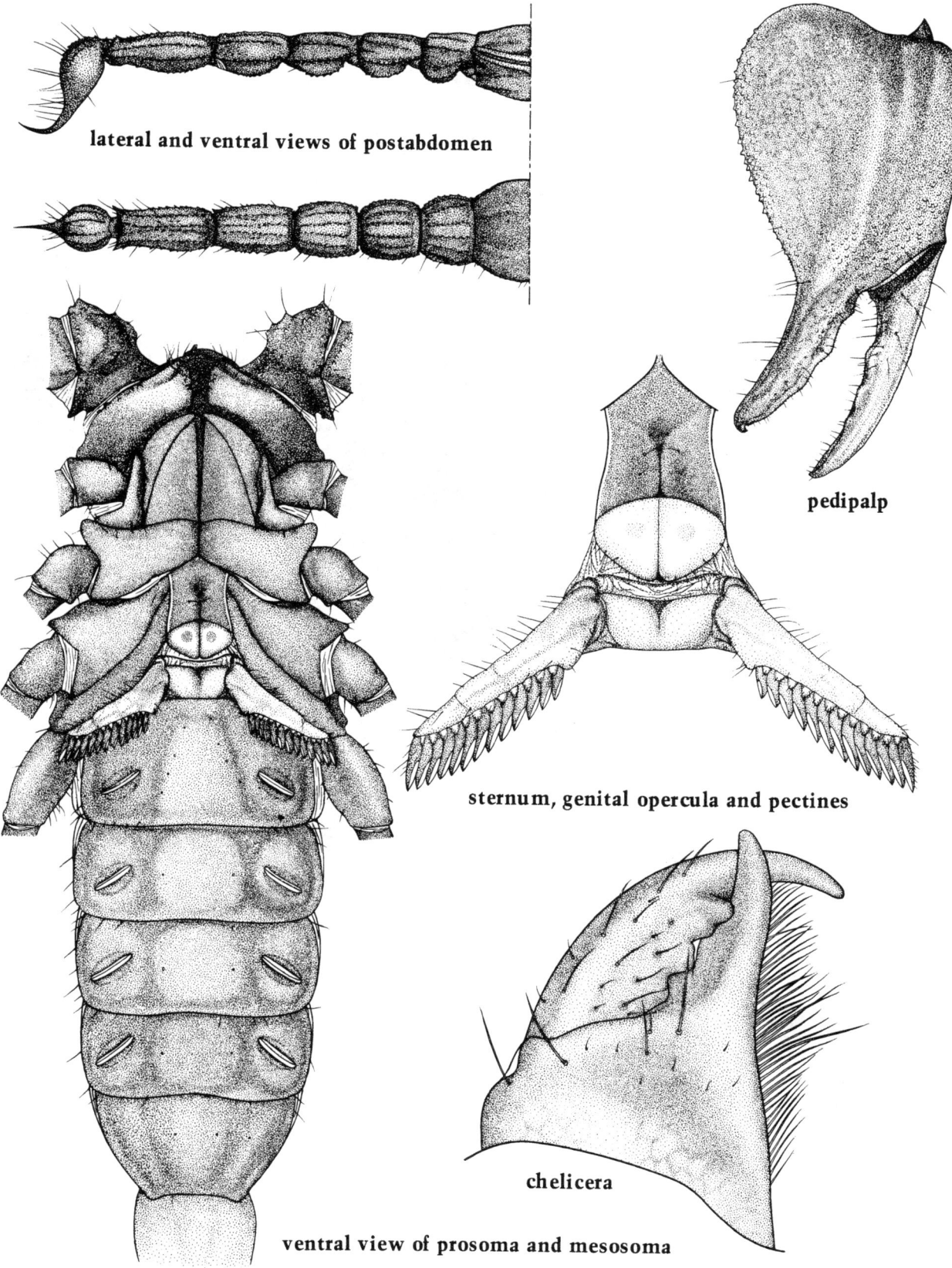

lateral and ventral views of postabdomen

pedipalp

sternum, genital opercula and pectines

chelicera

ventral view of prosoma and mesosoma

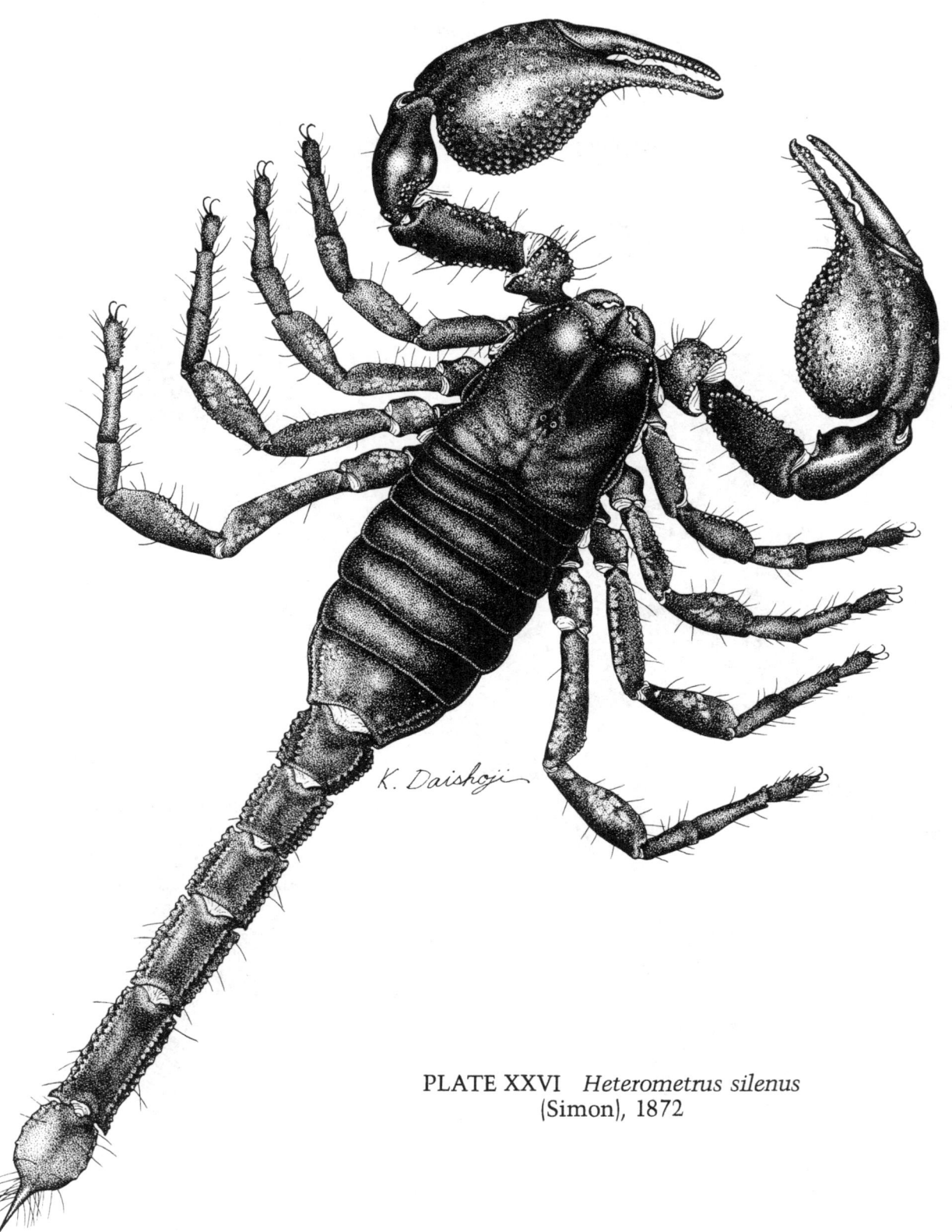

K. Daishoji

PLATE XXVI *Heterometrus silenus* (Simon), 1872

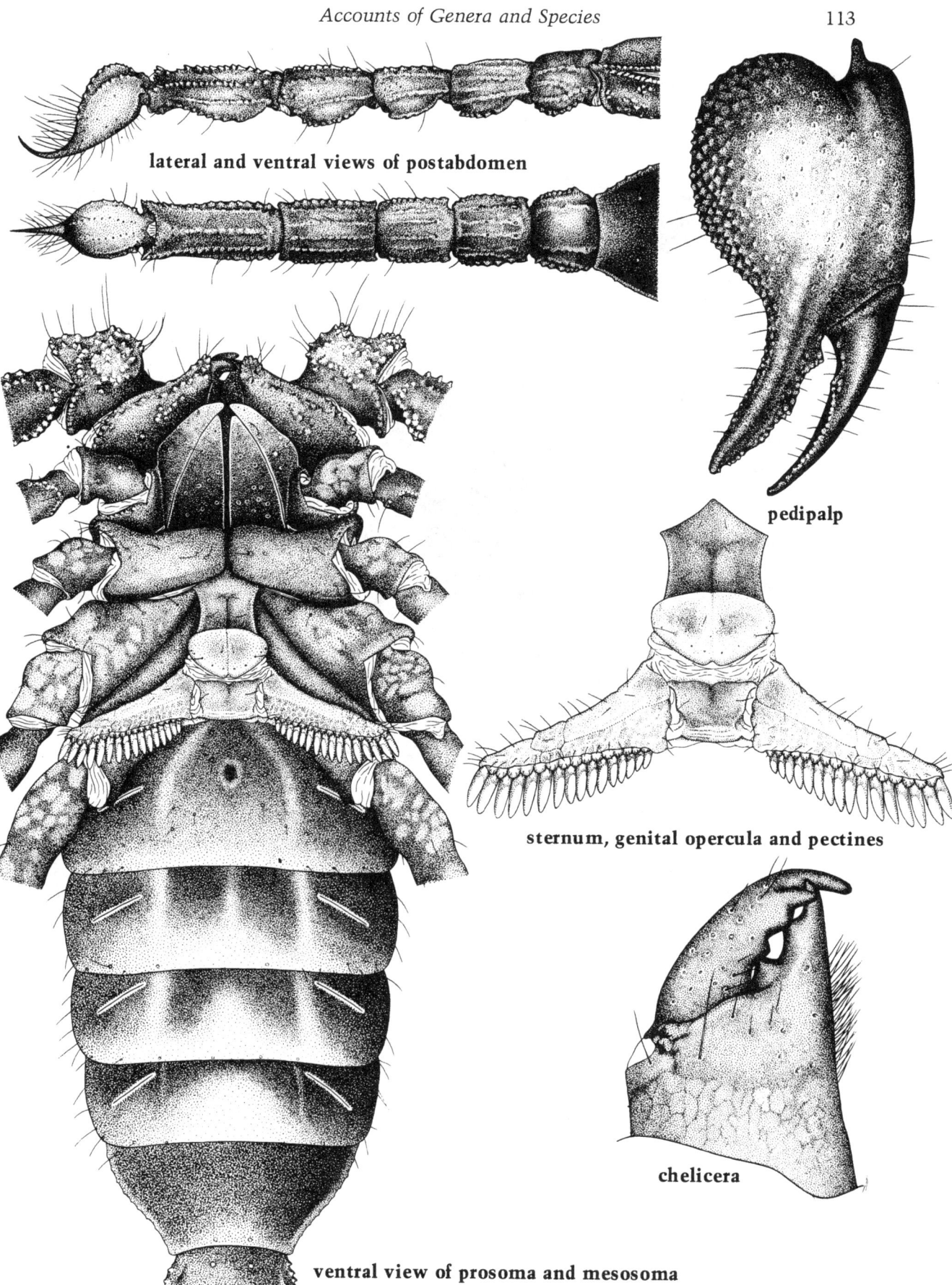

lateral and ventral views of postabdomen

pedipalp

sternum, genital opercula and pectines

chelicera

ventral view of prosoma and mesosoma

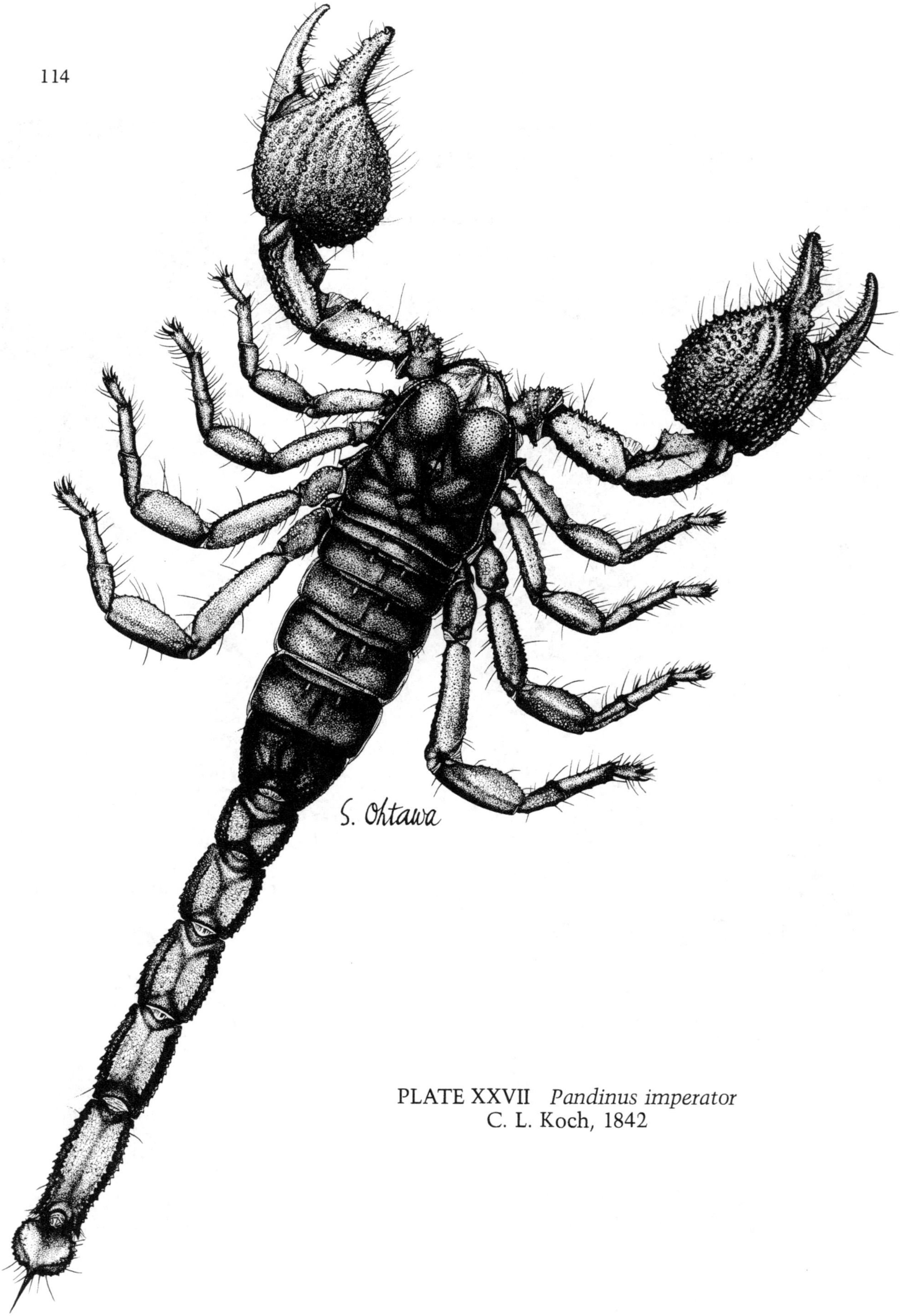

S. Ohtawa

PLATE XXVII *Pandinus imperator*
C. L. Koch, 1842

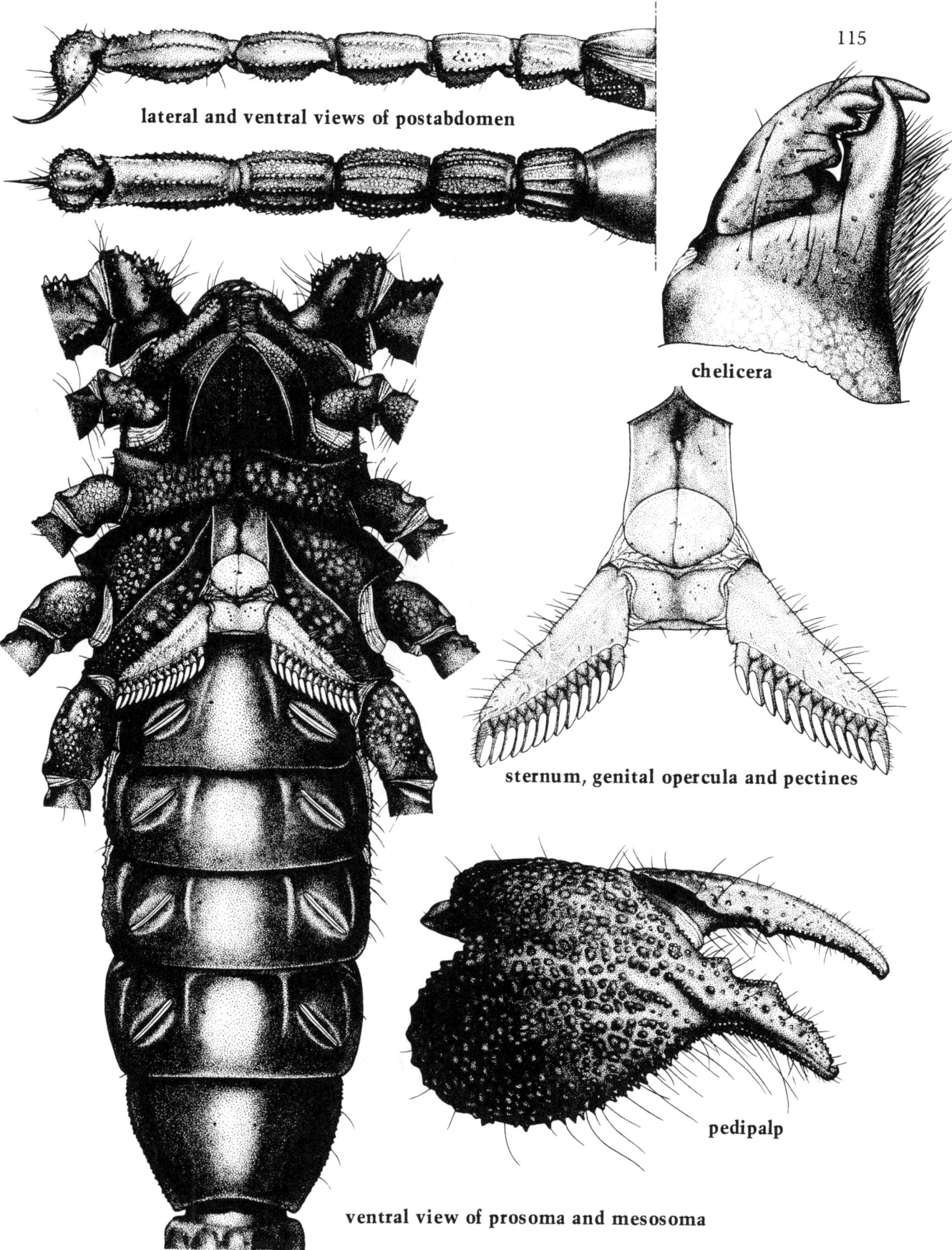

lateral and ventral views of postabdomen

chelicera

sternum, genital opercula and pectines

pedipalp

ventral view of prosoma and mesosoma

115

PLATE XXVIII *Pandinus (Pandinopsis) dictator*
(Pocock), 1888

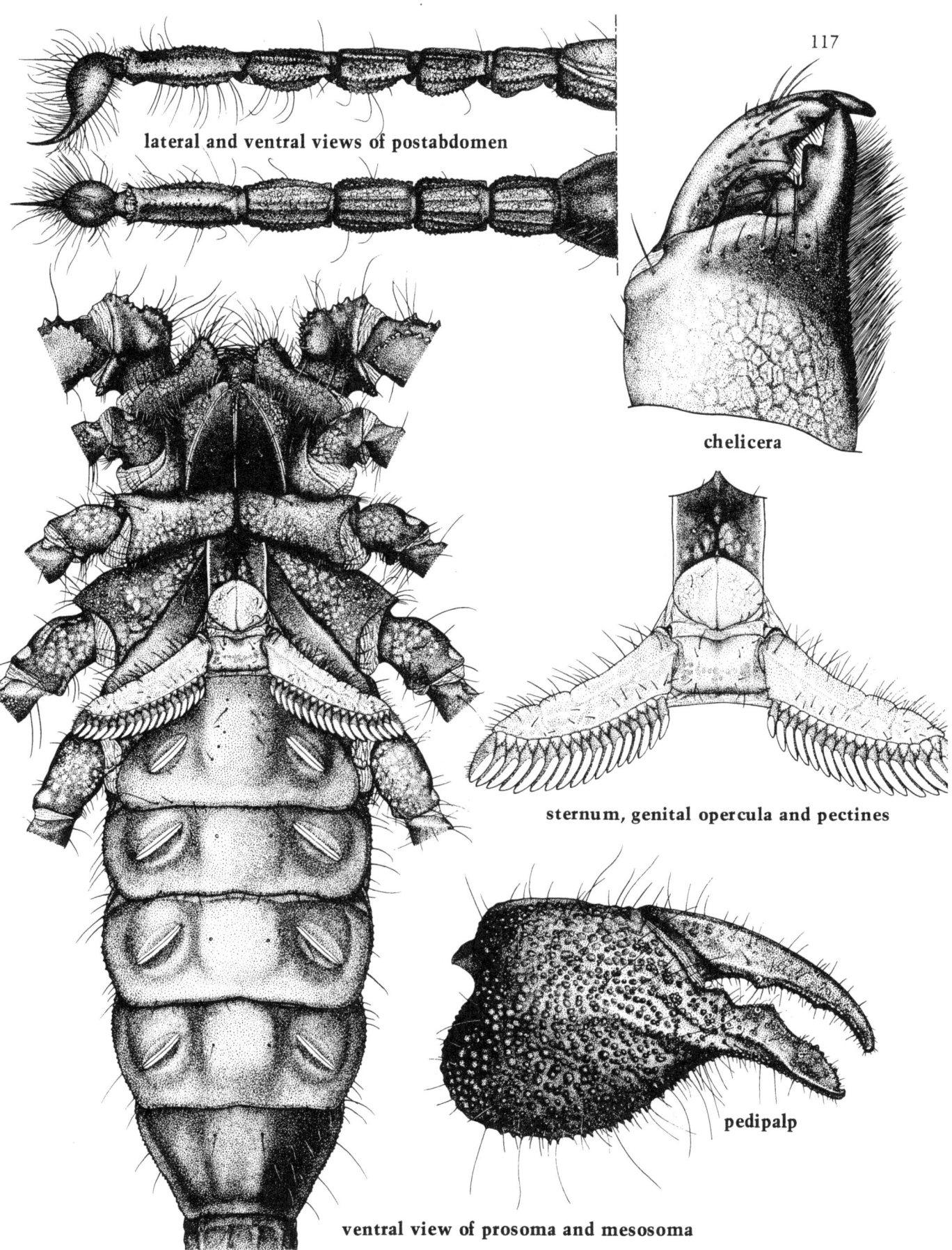

lateral and ventral views of postabdomen

117

chelicera

sternum, genital opercula and pectines

pedipalp

ventral view of prosoma and mesosoma

the species. This is probably due to its lack of medical importance. Kraepelin gave the distribution as West Africa (Camaroon to the Congo). The specimen figured was from Omboue, the Congo, collected in 1914 by C. R. Aschemeien. Its length was 172 mm.

Genus *Opisthophthalmus* C. L. Koch, 1838

Over 30 species of this genus occur in East and South Africa. A generic characteristic which serves to distinguish them from species of genus *Pandinus*, which they otherwise resemble, is the relatively straight anterior margin of the carapace. In addition, the median eyes are distinctly in the posterior half of the carapace. Although examples of many of the species attain a length of more than 100 mm, none is regarded as of significant medical importance. A description of the "courtship" and mating behavior of one species, *O. latimanus*, by Alexander (1957) indicated that unlike species of *Pandinus* and *Heterometrus*, it stridulates by rubbing its chelicerae on a striated area on the carapace. An example of only one species, *O. glabifrons* Peters, 1861, was on hand for illustration.

Opisthophthalmus glabifrons Peters, 1861 [PLATE XXIX]

Although the preserved specimen available for study was uniformly brown, Probst (1973) wrote that in life the trunk is yellow brown to rust brown, and the telson, legs, pedipalps, and anterior portion of the carapace yellow brown. The species has been reported in South Africa from Natal and the Transvaal, and in East Africa from Zambia, Malawi, and Tanzania. The specimen figured was seen through the courtesy of Dr. R. F. Lawrence, Director, Natal Museum, Pietermaritzburg, South Africa. This adult female from Natal was 87 mm in length.

Genus *Opisthacanthus* Peters, 1861

Distribution of species of this genus is unique in that while seven of the nine (theoretically) valid species occur in Africa, according to Mello-Leitao (1945) two, *O. cayaporum* Vellard, 1932 and *O. elatus* (Gervais, 1844), are found in the New World in Central and South America. While some species of the genus in Africa are found in areas where species of *Pandinus* and *Opisthophthalmus* also occur, there should be little difficulty in separation of species of these genera. Although the anterior margin of the carapace in species of this genus is deeply indented, as in *Pandinus*, the end of the tarsus is without rounded lobes. In addition, species of the genus are unique in that there are seldom more than 10 pectinal teeth.

There is no evidence that any of the species of the genus are of significant medical importance.

This relatively innocuous species is mentioned and figured here only because it is the sole member of the genus to occur in the United States. It has been found in Florida and also in Panama, Haiti, and Colombia. The species is sometimes referred to as *O. kinbergi* or *O. lepturus.* The specimen figured was collected at Ft. Sherman, Canal Zone, on 20 April 1925. Other specimens on hand for study were from Pearl Island, Panama and from the Empire Range, Canal Zone.

Hemiscorpion lepturus (Peters, 1861), a scorpion reported by Vachon (1966) to occur in Iraq and Iran, and a member of a genus with three other species found in the Mid East, was named by Pringle (1960) as a cause of both local and systemic effects of envenomation in Iraq. Bouisset and Larrouy (1962) wrote that while *Scorpion maurus* Linneus, 1758, a species widely distributed in southern Europe, the Middle East and North Africa, caused only "benign" effects by its sting in France, fatalities due to this species had been reported in Algeria. As the species has never been clearly defined, it may be that some of the "subspecies" found in Africa may possess venoms far more potent than that of the *S. maurus* recognized in Europe. While scorpion sting is not an important medical problem in Australia, Southcott (1976) wrote that stings by an undetermined species of *Urodacus* there had caused local swelling, redness, pain, and pyrexia in soldiers stung while on bivouac.

Although none of the species of the genus *Hadrurus* is of other than minor medical importance, large size, plus a wide distribution over the western United States and Mexico, makes them conspicuous members of the North American scorpion fauna. Stahnke (1969) has published a review of the taxonomy of the genus. Scorpions of this genus *Hadrurus* are frequently large (up to 11 cm in length), and possess an abundance of bristles on the legs and cauda. The pedal spurs are armed with denticulate spines. An important generic characteristic is the existence of a strong, spine-like tooth located ventrally on the movable finger of the chelicera. The carapace length approximates the width of the posterior portion. The median eyes

Opisthacanthus elatus
(Gervais, 1844)
[PLATE XXX]

Miscellaneous Species of Possible Medical Importance

FAMILY VEJOVIDAE

Genus *Hadrurus*
Thorell, 1876

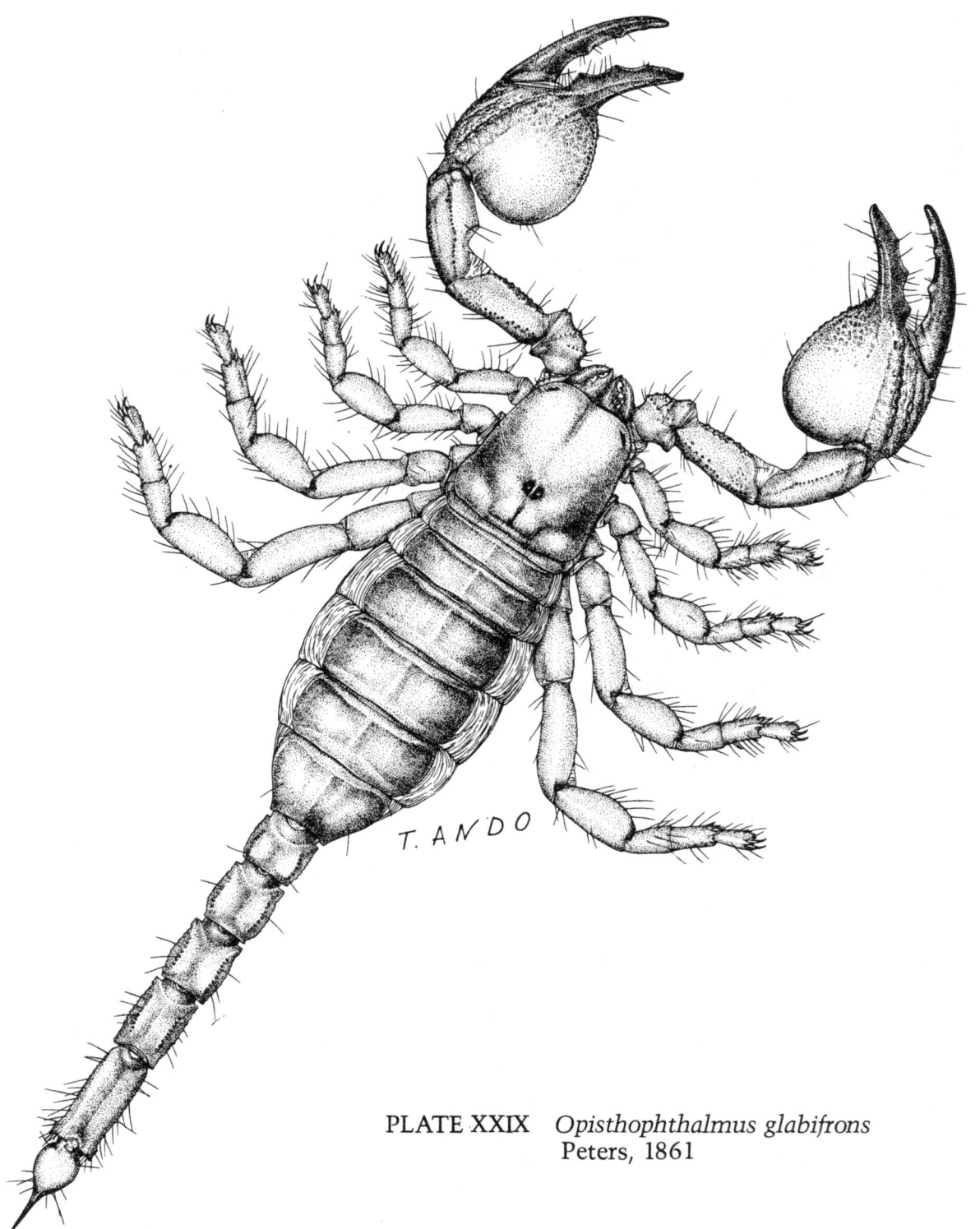

PLATE XXIX *Opisthophthalmus glabifrons*
Peters, 1861

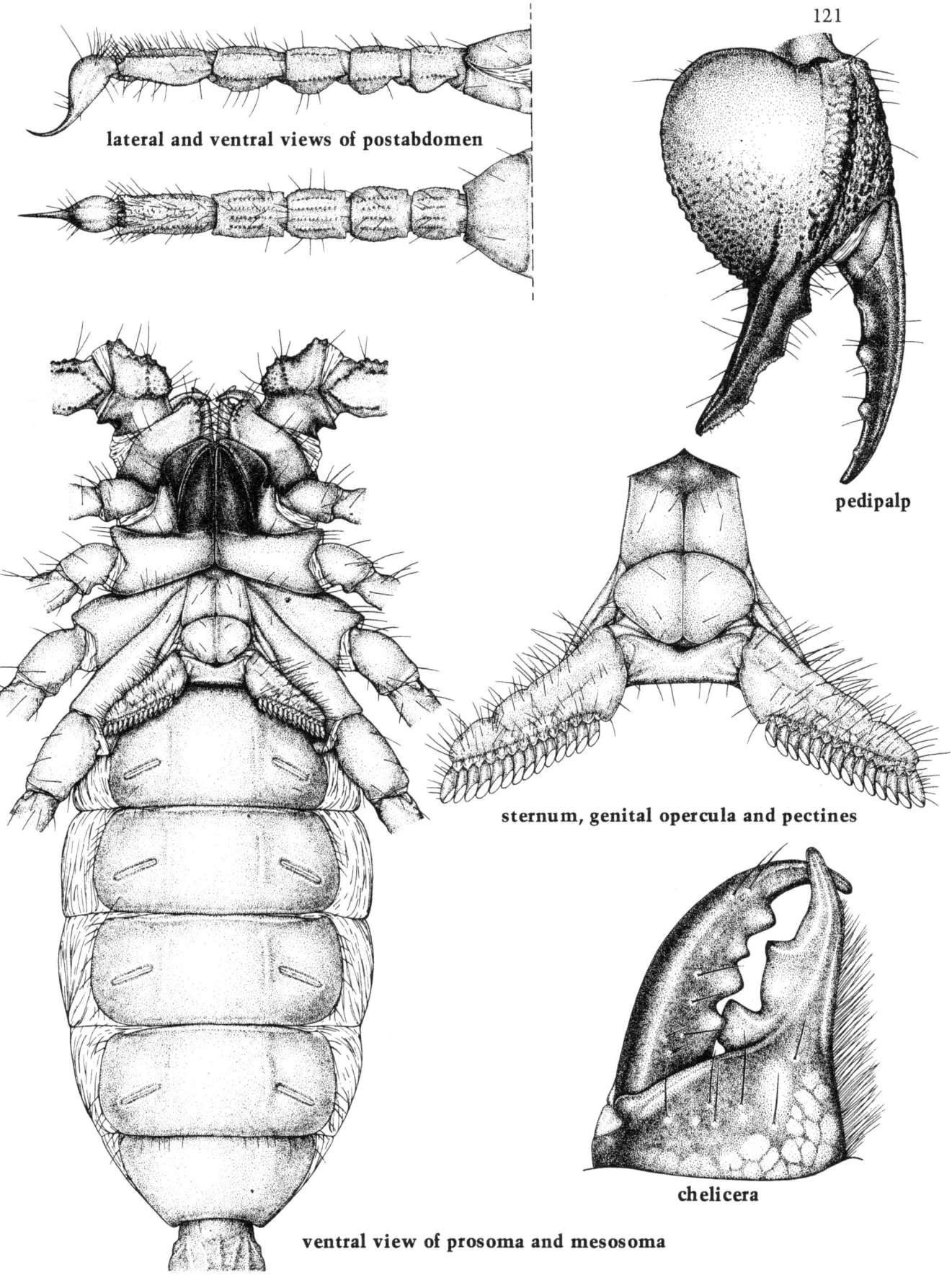

lateral and ventral views of postabdomen

121

pedipalp

sternum, genital opercula and pectines

chelicera

ventral view of prosoma and mesosoma

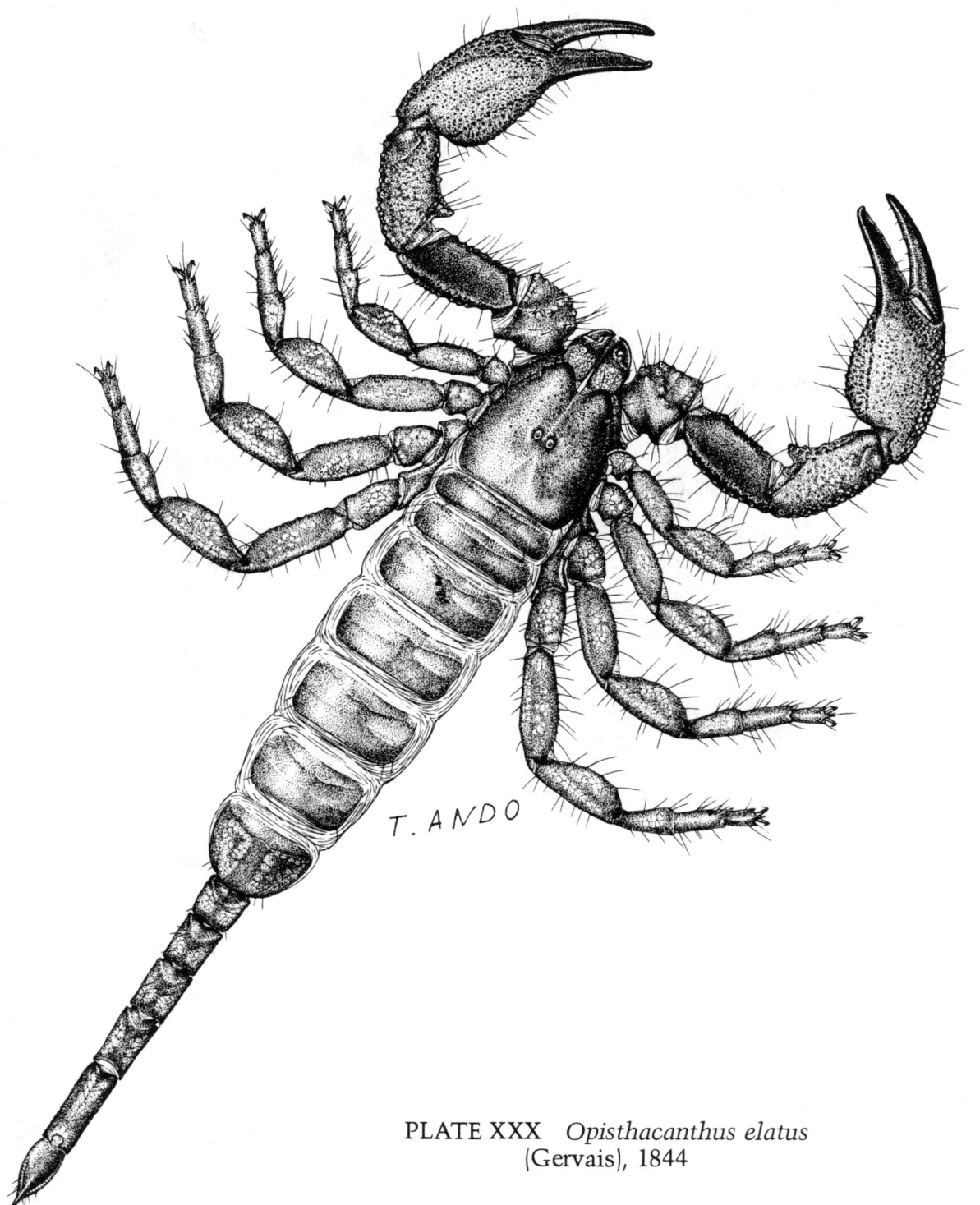

PLATE XXX *Opisthacanthus elatus*
(Gervais), 1844

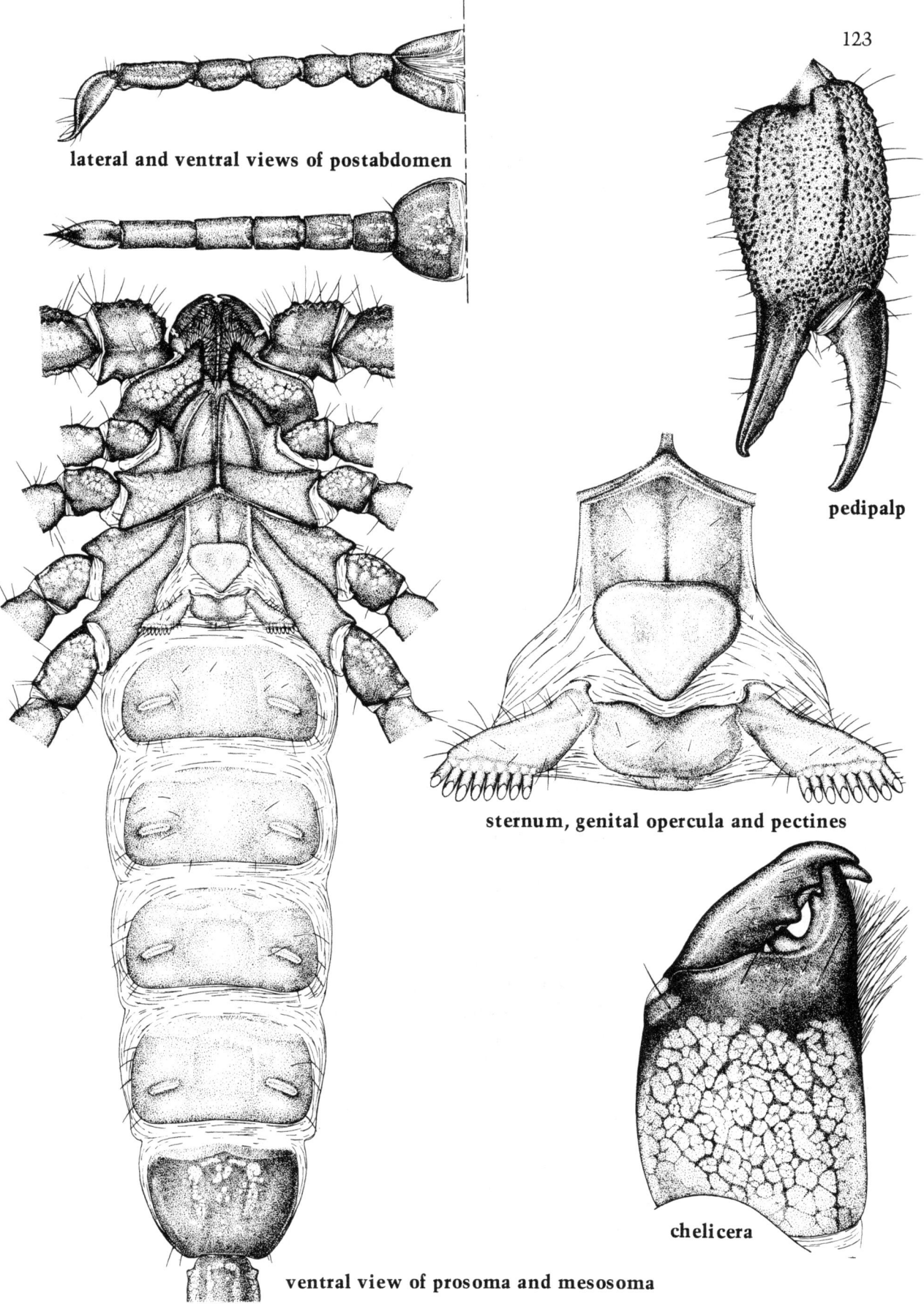

123

lateral and ventral views of postabdomen

pedipalp

sternum, genital opercula and pectines

chelicera

ventral view of prosoma and mesosoma

are slightly forward of the middle of the carapace. There are over 20 trichobothria on the ventral edge of the hand of the pedipalp and on the ventral and lateral surfaces of its patella. Like other members of the Vejovidae, species of this genus are "ground" scorpions and are found beneath logs, rocks and debris, often in burrows. Some, particularly in desert areas, may actively dig down to moisture during dry spells. Because of these habits, they are far less likely to enter homes than are the more dangerous buthid scorpions of genus *Centruroides* which are found in the same geographic areas. In his 1969 review, Stahnke recognized seven species.

Hadrurus hirsutus
(Wood, 1863)
[PLATE XXXI]

This scorpion, commonly known as the "giant hairy scorpion," is the largest of the genus and sometimes attains lengths of over 11 cm. The color of living specimens has been variously described as light yellow, dark yellow with a greenish hue on the carapace and preabdomen, and reddish brown. Ewing (1928) mentioned that the region in front of the ocular tubercle was white. Stahnke (1969) wrote of dark, crescent-shaped spots passing through the median ocular tubercle and extending to the lateral eyes. In the two preserved specimens on hand for study, the carapace and preabdomen were dark grey and the postabdomen and appendages were yellow.

H. hirsutus, like other members of the genus, is not a dangerously venomous species. Stahnke (1956) wrote that stings by *hirsutus* and the related species, *H. arizonensis* and *H. spadix* might result in localized, discolored swelling and sometimes severe burning pain which might radiate some distance from the site of the sting. Ennik (1972) mentioned that Russell, in a personal communication, stated that in his experience, stings by *Hadrurus* usually produced localized pain and some swelling and discoloration around the wound site.

Stahnke (1969) gave the distribution of the species as southwestern Arizona, southern California and Mexico. The specimen figured was determined as *hirsutus* by H. E. Ewing and had been collected in California by W. D. Pierce. The specific date and locality of collection were not given. The specimen was 87 mm in length.

Miscellaneous Studies
on the Vejovidae

Published reports indicate that stings by vejovid scorpions other than species of *Hadrurus* produce much the same effect in man. Russell *et al* (1968) wrote that stings by *Vejovis spinigerus*, a species occurring in the southwestern United

States, produce transient pain with occasional swelling, localized paresthesia and tenderness. In some cases skin temperature tends to be elevated. In more severe cases there may be some weakness of the involved part and, occasionally, muscle fasciculations. Symptoms and signs usually disappear within 24 hours. Williams (1970) reported similar effects from a sting by *Vejovis confusus,* a species found in the Sonoran Desert of Arizona. The sting was followed by intense pain, numbness and swelling which prevented effective use of the fingers and wrist for 24 hours.

An early, but still valuable, paper concerning the vejovids and other scorpions of the western United States is that of Ewing (1928). The recent monograph by Gertsch and Soleglad (1972) gives a thorough review of the North American vejovid scorpions of genus *Uroctonus* as well as descriptions of several new species of genus *Vejovis,* and redescriptions of others. This paper is an outstanding example of utilization of modern taxonomic techniques and is of particular interest because of its demonstration of the importance of trichobothrial patterns in classification.

An excellent study of the birth and postbirth activities of several species of vejovid scorpions was published by Williams (1969). In all scorpions of this family studied, the newly born young arrange themselves in an orderly fashion on the back of the mother, side by side and facing anteriorly. This is in marked contrast to the situation in the Buthidae where the young are arranged at random, sometimes three or four deep, with no particular directional orientation. Early growth rates in offspring of *Vejovis spinigerus* were described by McAlister (1960).

Although the distribution, habits and methods of control of dangerously venomous scorpions are quite well known, it is by no means certain that this information will lead to a marked decrease in numbers of scorpion stings. In many countries where scorpion envenomation is a problem, the cost of "scorpion-proofing" houses and of application of insecticides is a prohibitive expense.

Similarly, while research on chemical and pharmacological properties of scorpion venoms has greatly increased our knowledge in this field, there have been no recent dramatic improvements in treatment of envenomation by scorpions. Specific antivenin remains the treatment of choice in countries where the problem is of sufficient magnitude as to war-

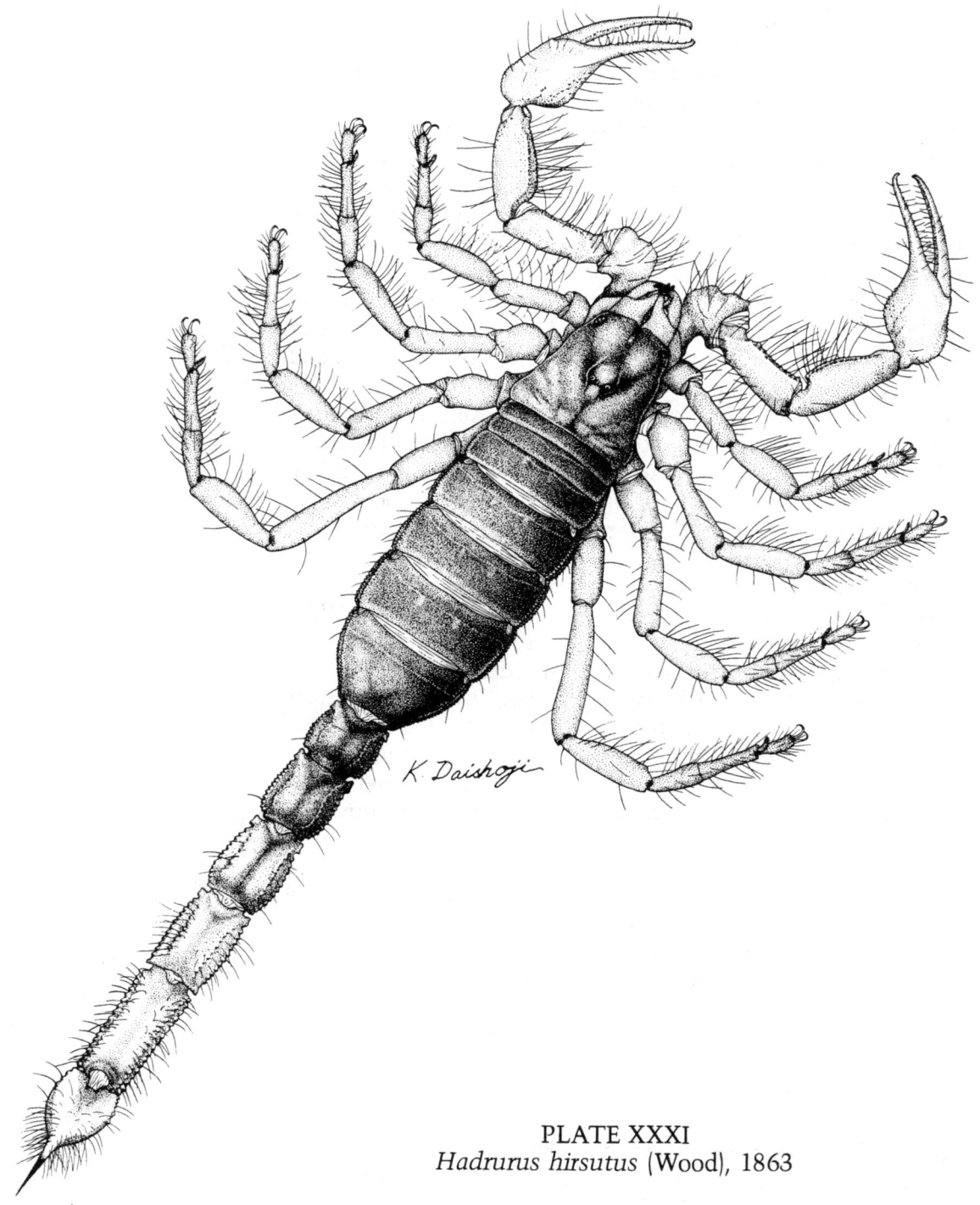

PLATE XXXI
Hadrurus hirsutus (Wood), 1863

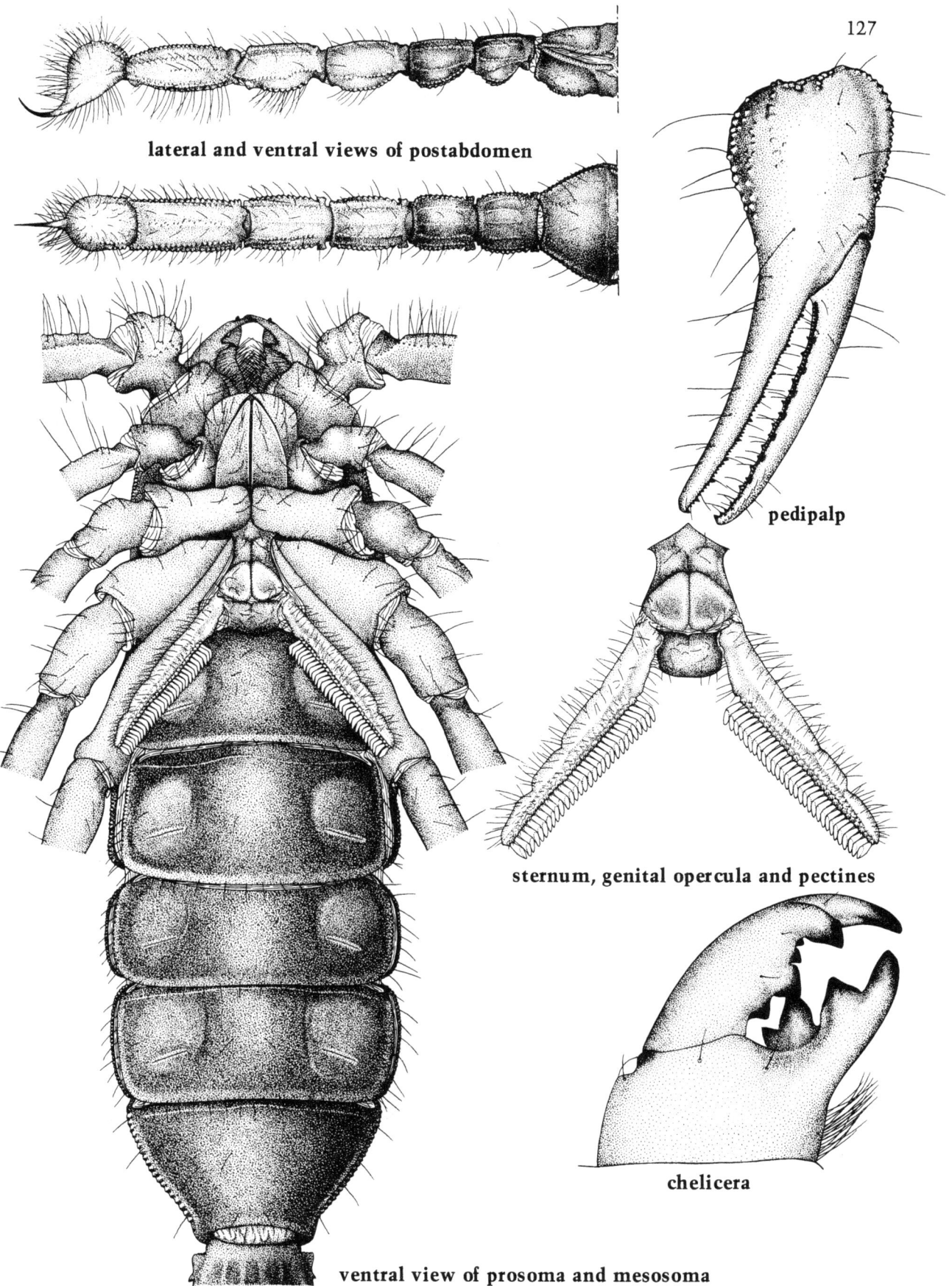

127

lateral and ventral views of postabdomen

pedipalp

sternum, genital opercula and pectines

chelicera

ventral view of prosoma and mesosoma

rant antivenin production. Even in those areas, however, availability of antivenin and of skilled medical attention is often not adequate for treatment of victims living in remote areas. This is of particular importance since many persons may show an allergic response to the antivenin, which is usually prepared through hyperimmunization of horses, and which therefore should be administered only by trained medical or paramedical personnel.

It is unfortunate that most scorpion antivenins now available, with a few exceptions, are rather narrowly specific in effectiveness and useful only in treatment of stings by scorpions whose venoms were used in preparation of the product. Development of a polyvalent antivenin effective in most, if not all, regions where dangerously venomous scorpions exist would be desirable, but efforts toward that goal, up to the present, have not been successful.

REFERENCES CONCERNING GENUS ANDROCTONUS

Balozet, L. 1964. Le Scorpionisme en Afrique de Nord. *Bull. Soc. Pathol. Exot.*, *57:* 37–38.

———. 1971. Scorpionism in the Old World. Ch. 56 *In: Venomous Animals and Their Venoms*, vol. III, *Venomous Invertebrates*. Bücherl, W. and Buckley, E. E., eds. Academic Press, New York, pp. 349–371.

Levy, M. 1965. Les envenimations par piqûres de scorpions à Marrakech. (Maroc). Thèse, Faculté de Médicine, Paris, 42 pp.

Minning, W. and Zumpt, F. 1942. Scorpione Nordafrikas. Merkblätter des Instituts für Schiffs and Tropenkrankeiten/Hamberg, Mediziniech wichtige Spinnentiere. Merkblatt 1.

Pocock, R. I. 1900. The Fauna of British India including Ceylon and Burma. *Arachnida.* Taylor and Francis, London, 279 pp.

Rolli, K. 1972. Essais de differents insecticides dans la destruction des scorpions. *Arch. de L'Institut Pasteur de Tunis*, *49:* 267–274.

Shulov, A. 1939. The venom of the scorpion *Buthus quinquestriatus* and the preparation of an anti-serum. *Trans. Royal Soc. Trop. Med. and Hyg.*, XXIII, 253–256.

———. 1962. On some Israeli scorpions. Dapin Refuiim (Folia Medica), XXI:3–14.

Tulga, T. 1960. Cross-reactions between anti-scorpion *(Buthus quinquestriatus)* and anti-scorpion *(Prionurus crassicauda)* sera. *Turk. Hig. Tecr. Biyol. Derg.*, *20:* 191–203.

———. 1964. Scorpions found in Turkey and paraspecific action of an antivenin produced with the venom of the species: *Androctonus crassicauda. Turk. Hig. Tecr. Biyol. Derg.*, *24:* 153–155.

Vachon, M. 1952. *Etudes sur Les Scorpions*. Institut Pasteur d'Algerie, Lager, 482 pp.

————. 1966. Liste des scorpions connus en Egypte, Arabie, Israel, Liban, Syrie, Jordanie, Turquie, Irak, Iran. *Toxicon, 5:* 209–218.

Whittemore, F. W., Jr., Keegan, H. L. and Borowitz, J. L. 1961. Studies of scorpion antivenins. 1. Paraspecificity. *Bull. Wld. Hlth. Org., 25:* 185–188.

Balozet, L. 1971. Scorpionism in the Old World. Ch. 56 *In: Venomous Animals and Their Venoms*, vol. III, *Venomous Invertebrates.* Bücherl, W. and Buckley, E. E., eds. Academic Press, New York, pp. 349–371.

Deoras, P. J. 1961. A study of scorpions: Their distribution, incidence, and control in Maharashtra. *Probe, 1*: 45–54.

Mundle, P. M. 1961. Scorpion stings. *Brit. Med. J., 1*: 1042.

Pocock, R. I. 1900. *The Fauna of British India.* Taylor and Frances, London, 379 pp.

Reddy, C. R. R. M., Suvarnakum, G., Devi, C. C. and Reddy, C. N. 1972. Pathology of scorpion venom poisoning. *J. Trop. Med. and Hyg., 75:* 98–100.

Santhanakrishnan, B. R. and Balagopal Raju, V. 1974. Management of scorpion sting in children. *J. Trop. Med. Hyg., 77:* 133–135.

Vachon, M. and Stockmann, R. 1968. Contributions à l'étude des scorpions africains appartenant au genre *Buthotus* Vachon 1949 et étude de la variabilité. *Monitore Zool. Ital.* (N.S.) 2 (Suppl.): 81–149.

Whittemore, F. W., Keegan, H. L., Fitzgerald, C. M., Bryant, H. A., and Flanigan, J. F. 1963. Studies of scorpion antivenins. 2. Venom collection and scorpion colony maintenance. *Bull. Wld. Hlth. Org., 28:* 505–511.

Balozet, L. 1971. Scorpionism in the Old World. Ch. 56 *In: Venomous Animals and Their Venoms*, vol. III, *Venomous Invertebrates*, Bücherl, W. and Buckley, E. E., eds. Academic Press, New York, pp. 349–371.

Bouisset, L. and Larrouy, G. 1962. Envenimations par *Scorpio maurus* et *Buthus occitanus* dans le Department de Tlemcen. *Bull. Soc. Path. Exot., 55*: 139–146.

Fabre, J. H. 1911. The Languedocian Scorpion. Chaps. 17 & 18 *In: The Life and Love of the Insect.* A. & C. Black, Ltd., London, pp. 223–260.

Kraepelin, K. 1899. Scorpiones und Pedipalpi. Das Tierreich. Friedlander und Sohn, Berlin. 8 Lieferung, 265 pp.

Rolli, K. 1972. Éssais de differents insecticides dans la destruction des scorpions. *Arch. de L'Institut Pasteur de Tunis*, 49: 267–274.

Vachon, M. 1952. *Études sur Les Scorpions.* Institut Pasteur d'Algerie, Lager, 482 pp.

Wahbeh, Y. 1965. Scorpion stings in children. *Jordan Med. J., 1:* 57–61

Whittemore, F. W., Jr., Keegan, H. L., and Borowitz, J. L. 1961. Studies of scorpion antivenins. 1. Paraspecificity. *Bull. Wld. Hlth. Org., 25:* 185–188.

Balozet, L. 1971. Scorpionism in the Old World. Ch. 56 *In: Venomous Animals and Their Venoms*, vol. III, *Venomous Invertebrates.* Bücherl, W. and Buckley, E. E., eds. Academic Press, New York, pp. 349–371.

REFERENCES
CONCERNING
GENUS BUTHOTUS

REFERENCES
CONCERNING
GENUS BUTHUS

REFERENCES
CONCERNING
GENUS LEIURUS

Minning, W. and Zumpt, F. 1942. Scorpione Nordafrikas. Merkblätter des instituts für Schiffs und Tropenkrankeiten/Hamberg, Mediziniech wichtige Spinnentiere. Merkblatt 1.

Shulov, A. 1962. On some Israeli scorpions. *Dapin Refuiim* (Folia Medica), XXI: 3–14.

Tulga, T. 1960. Cross-reactions between anti-scorpion (*Buthus quinquestriatus*) and anti-scorpion (*Prionurus crassicauda*) sera. *Turk. Hig. Tecr. Biyol. Derg.*, 20: 191–203.

———. 1964 Scorpions found in Turkey and paraspecific action of an antivenin produced with the venom of the species: *Androctonus crassicauda. Turk. Hig. Tecr. Biyol. Derg.*, 24: 153–155.

Vachon, M. 1952. *Études sur Les Scorpions.* Institut Pasteur d'Algérie, Lager, 482 pp.

Whittemore, F. W. and Keegan, H. L. 1963. Medically important scorpions in the Pacific area. *In: Venomous and Poisonous Animals and Noxious Plants of the Pacific Region.* Keegan, H. L. and Macfarlane, W. V., eds. Pergamon Press, New York, pp. 107–110.

Whittemore, F. W., Keegan, H. L., Fitzgerald, C. M., Bryant, H. A., and Flanigan, J. F. 1963. Studies of scorpion antivenins. 2. Venom collection and scorpion colony maintenance. *Bull. Wld. Hlth. Org., 28:* 505–511.

REFERENCES CONCERNING GENUS PARABUTHUS

Hewitt, J. 1918. A survey of the scorpion fauna of South Africa. *Trans. Roy. Soc. S. Africa. 6:* 89–192.

Probst, P. J. 1976. A review of the scorpions of East Africa with special regard to Kenya and Tanzania. *Acta Tropica, 30:* 312–334.

Stahnke, H. L. 1972. A key to the genera of Buthidae (Scorpionida). *Ent. News, 81:* 297–316.

Vachon, M. 1966. Liste des scorpions connus en Egypte, Arabie, Israel, Liban, Syrie, Jordanie, Turquie, Irak, Iran. *Toxicon.,* 4: 209–218.

REFERENCES CONCERNING GENUS TITYUS

Abalos, J. W. 1963. Scorpions of Argentina. *In: Venomous and Poisonous Animals and Noxious Plants of the Pacific Region.* Keegan, H. L. and Macfarlane, W. V., eds. Pergamon Press, Oxford, pp. 111-117

Bücherl, W. 1971. Classification, Biology, and Venom Extraction of Scorpions. Ch. 55 *In: Venomous Animals and Their Venoms,* vol. III, *Venomous Invertebrates.* Bücherl, W. and Buckley, E. E., eds. Academic Press, New York, pp. 317–347.

———. 1956. Scorpions and their effects in Brazil. IV. Observations on insecticides lethal to scorpions and other methods of control. Mem. Inst. Butantan, 27: 107–120.

Ewing, H. E. 1928. The scorpions of the western part of the United States with notes on those occurring in Northern Mexico. Proc. U.S. Nat. Mus., 73: 1–24.

Floch, H., Barrat, R. and Abonnenc, E. 1950. L'envenimation par piqure de scorpions en Guyane francaise. Inst. Past. et Territoire Inini. Publ., 219: 1–4.

Lourencõ, W. R. Nocoes sobre aracnidismo. Cerrado, no. 30, December, 1975, 25–28. (This is a publication of the Zoobotanical Foundation of the Secretariat of Agriculture and Industry of the Government of the Distrito Federal, Brazil).

Matthiesen, F. A. 1962. Parthenogenesis in Scorpions. *Evolution. 16:* 255–256.

Mello-Leitao, C. 1945. Escorpiões Sul-Americanos. Arq. Museu Nac., XL. 468 pp., Rio de Janeiro.

Muma, M. H. 1967. Arthropods of Florida and Neighboring Land Areas. Vol. 4, (Publication of the Florida Dept. of Agriculture) 28 pp.

Poon-King, T. 1963. Myocarditis from scorpion stings. *Brit. Med. J., 1:* 374–377.

San Martin, P. R., and Gambardella, L. A. de. 1966. Nueva comprobacion de la partenogenesis en *Tityus serrulatus* Lutz and Mello-Campos, 1922 (Scorpionida, Buthidae), *Rev. Soc. Ent. Arg. XXVIII:* 79–84.

Souza, J. C. de, Bustamente, F. M. de and Bicalho, J. C. 1954. Novos dados sobre o combate aos escorpioes em belo horizonte com o hexa-chlorociclohexona. *Rev. Brazil de Malariol. e Doencas Tropicais, 6:* 357–361.

Stahnke, H. L. 1972. A key to the genera of Buthidae (Scorpionida). *Ent. News, 81:* 297–316.

Waterman, J. A. 1938a. Some notes on scorpion poisoning in Trinidad. *Trans. Roy. Soc. Trop. Med. & Hyg., 31:* 607–624.

———. 1938b. A few cases. *Caribbean Med. J., 1:* 119–120.

———. 1950a. Two cases of scorpion poisoning characterized by convulsions with electrocardiograms. *Caribbean Med. J., XII:* 127–129.

———. 1950b. Scorpions in the West Indies with special reference to *Tityus trinitatus. Caribbean Med. J., XII:* 167–177.

REFERENCES CONCERNING GENUS CENTRUROIDES

Baerg, W. J. 1961. Scorpions: Biology and Effect of their Venom. Bull. 649 Agriculture Experiment Station. Div. of Agriculture, U. of Arkansas, pp. 1–34.

Díaz Nájera, A. 1964 Alacranes de la republica Mexicana: Identification de ejemplarea capturados en 235 localidades. *Rev. Inst. Salubr. Enferm. trop.* (Mex). XXIV: 15–30.

Ennik, F. 1972. A short review of scorpion biology, management of stings, and control. *California Vector Views.* 19: 69–80.

Ewing, H. E. 1928. The scorpions of the western part of the United States with notes on those occurring in northern Mexico. Proc. U. S. Nat. Mus., 73: 1–24.

Hoffman, C. C. 1932. Monografias para la Entomologia Medica de Mexico. Monografia Num. 2. Los Escorpiones de Mexico. Segunda parte. Buthidae. An. Inst. Biol. 3: 243–361.

Keegan, H. L. and Lockwood, W. R. 1971. Secretory epithelium in venom glands of two species of scorpion of the genus *Centruroides* Marx. *Amer. J.Trop. Med. & Hyg.,* 20: 770–785.

Marinkelle, C. J. and Stahnke, H. L. 1965. Toxicological and clinical studies on *Centruroides margaritatus* (Gervais), a common scorpion in western Colombia. *J. Med. Ent.,* 2: 197–199.

Mazzotti L. and Bravo-Becherelle. 1963. Scorpionism in the Mexican Republic. *In: Venomous and Poisonous Animals and Noxious Plants of the Pacific Region.* A collection of papers based on a symposium in the public health and medical science division at the Tenth Pacific Sci-

132 SCORPIONS OF MEDICAL IMPORTANCE

ence Congress, H. L. Keegan and W. V. Macfarlane, eds. Pergamon Press, New York, pp. 119–131.

Muma, M. H. 1967. Scorpions, Whip Scorpions and Wind Scorpions of Florida. Volume 4 in the series Arthropods of Florida. Florida Dept. of Agriculture, pp. 1–28.

Smith, F. R. 1927. Observations on scorpions. *Science*, 65: 64.

Stahnke, H.L. 1956. Scorpions. Revised edition. Poisonous Animals Research Laboratory. Arizona State College, Tempe, Arizona. 36 pp.

———. 1966. Some aspects of scorpion behavior. *Bull. So. Calif. Acad. Sci.*, *65:* 65–79.

———. 1971. Some observations of the genus *Centruroides* (Buthidae, Scorpionida). *Ent. News. 82:* 281:307.

Stahnke, H. L. and Calos M. 1977. A key to the species of the genus *Centruroides* Marx (Scorpionida: Buthidae). *Ent. News, 88:* 111–120.

Whittemore, F. W. and Keegan, H. L. 1963 Medically important scorpions in the Pacific area. *In: Venomous and Poisonous Animals and Noxious Plants of the Pacific Region.* A collection of papers based on a symposium in the public health and medical science division at the Tenth Pacific Science Congress. H. L. Keegan and W. V. Macfarlane, eds. Pergamon Press, New York, pp. 107–109.

Whittemore, F. W., Keegan, H. L., Fitzgerald, C. M., Bryant, H. A., and Flanigan, J. F. 1963. Studies of scorpion antivenins: 2. Venom collection and scorpion colony maintenance. *Bull. Wld. Hlth. Org., 28:* 505–511.

REFERENCES CONCERNING FAMILY SCORPIONIDAE

Alexander, A. J. 1957. The courtship and mating of the scorpion, *Opisthophthalmus latimanus. Proc. Zool. Soc. Lond., XX 128:* 529–544.

———. 1958. On the stridulation of scorpions. *Behavior.* 12: 339–352.

Baerg, W. J., 1961. Scorpions: *Biology and Effect of Their Venom.* Agriculture Experiment Station, University of Arkansas, Fayetteville, Bull. 649. 34 pp.

Balozet, L., 1971. Scorpionism in the Old World. Ch. 56 *In: Venomous Animals and Their Venoms,* vol. III, *Venomous Invertebrates,* Bücherl, W. and Buckley, E. E., eds. Academic Press, New York, pp. 349–371.

Bouisset, L., and Larrouy, G., 1962. Envenimations par *Scorpio maurus et Buthus occitanus* dans le Department de Tlemcen (Algeria). *Bull. Soc. Path. Exot., 55:* 139–146.

Deoras, P. J., 1961. A study of scorpions. Their distribution, incidence and control in Maharashtra, *Probe, 1:* 45–54.

Kopstein, F., 1927. The poison of the Javanese giant scorpion *Heterometrus cyaneus. Meded. Dienst. d. Volksgezondheid in Nederl. Indie., 3:* 1–10.

———., 1932. Die giftiere Java's und ihre bedeutung für den Menschen. *Mededeel. v. d. dienst. f. volksgezondh. in Nederl. Indie., 21:* 222–256.

Kraepelin, K., 1899. *Scorpiones und Padipalpi. Das Tierreich.* Friedlander und Sohn, Berlin. 8 Lieferung, 265 pp.

Mello-Leitao, C. de., 1945. *Escorpioes Sul-Americanos.* Arquivos do Museu Nacion vol. XL, 468 pp.

Pocock, R. I., 1900a. *The Fauna of British India.* Taylor and Frances, London, 379 pp.

————., 1900b. The scorpions of the genus *Heterometrus. Ann. Nat. Hist.,* VI: 361–365.

Pringle, G. 1960. Notes on the scorpions of Iraq. *Bull. Endem. Dis., 3:* 73–87.

Probst, P. J., 1973. A review of the scorpions of East Africa with special regard to Kenya and Tanzania. *Acta Tropica, 30:* 312–335.

Schultze, W., 1927. Biology of the large Philippine forest scorpion. *Philip. J. Sci., 32:* 375–393.

Southcott, R. V., 1976. Arachnidism and allied syndromes in the Australia region. *Records of the Adelaide Children's Hospital. 1:* 97–186.

Vachon, M. 1966. Liste des scorpions cohnus en Egypte, Arabie, Israel, Liban, Syrie, Jordanie, Turquie, Irak, Iran. *Toxicon, 4:* 209–218.

————., 1967. Le grand Scorpion du Senegal: *Pandinus gambiensis* Pocock, 1899 doit etre considere comme une veritable espece et non comme un sous-espesce de *Pandinus imperator* C. L. Koch, 1842. *Bull. IFAN, ser. A., 29:* 1534–1537.

Vachon, M., Roy, R. and Condamin, M. 1970. Le developpment postembryonnaire du scorpion *Pandinus gambiensis* Pocock. *Bull. IFAN, ser. A, 32:* 412–432.

————., 1973. Etude des caracteres utilises pour classer les families et les genres de Scorpions (Arachnides). *Bull. Mus. HIst. nat. Paris. 3ᵉ serie, no. 140* (Zoologie), *104:* 857–958.

Whittemore, F. W., Keegan, H. L., Fitzgerald, C. M., Bryant, H. A., and Flanigan, J. F., 1963. Studies of scorpion antivenins. 2. Venom collection and scorpion colony maintenance. *Bull. Wld. Hlth. Org., 28:* 505–511.

Yoshida, Y. and Toshioka, S. 1964. Studies on spermatogenesis in scorpions. 1. Numbers of chromosomes in male germ-cells of three species of scorpions. *Acta Arachnologica. XIX:* 1–4.

REFERENCES CONCERNING FAMILY VEJOVIDAE

Ennik, F., 1972. A short review of scorpion biology, management of stings, and control. *Calif. Vector Views, 19:* 69–80.

Ewing, H. E., 1928. The scorpions of the western part of the United States, with notes on those occuring in northern Mexico. *Proc. U. S. Natl. Mus., 73:* 1–24.

Gertsch, W. J., and Soleglad, M. 1972. Studies of North American scorpions of the genera *Uroctonus* and *Vejovis* (Scorpionida, Vejovidae). *Bull. American Mus. Nat. Hist., 148:* 547–608.

McAlister, W. H., 1960. Early growth rates in offspring from two broods of *Vejovis spinigerus* Wood. *Texas J. Sci., XII:* 158–162.

Russell, R. E., Alender, C. B., and Buess, F. W., 1968. Venom of the scorpion *Vejovis spinigerus. Science, 159:* 90–91.

Stahnke, H. L., 1969. A review of *Hadrurus* scorpions (Vejovidae). *Ent. News, 80:* 57–65.

————., 1956. *Scorpions.* Published by Poisonous Animals Research Laboratory, Arizona State College, Tempe, Arizona, 37 pp.

Williams, S. C., 1969. Birth activities of some North American scorpions. *Proc. Calif. Acad. Sci., XXXVII:* 1–24.

————., 1970. The effects on man of a natural sting by the scorpion *Vejovis confusus* Stahnke. *Pan-Pacific Entomol., 46:* 77–78.

Subject Index

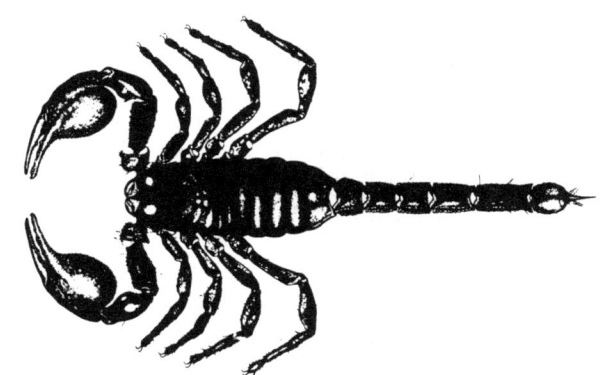

135

Index

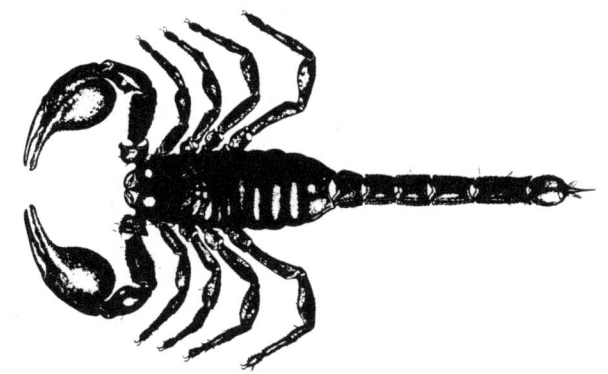